THE ECOLOGY OF EXPANSION AND ABANDONMENT

THE ECOLOGY OF EXPANSION AND ABANDONMENT

Medieval and post-medieval agriculture and settlement in a landscape perspective

Per Lagerås

National Heritage Board, Sweden

RIKSANTIKVARIEÄMBETET
Box 5405, SE-114 84 Stockholm, Sweden
Phone +46 (0)8-5191 8000
www.raa.se

Swedish customer service:
www.arkeologibocker.se
Phone +46 (0)31-334 29 05, +46 (0)31-334 29 04
Fax +46 (0)31-334 29 01

International customer service:
Oxbow Books, Park End Place, Oxford OX1 1HN, UK
www.oxbowbooks.com
Phone +44 (0)1865-241249
Fax +44 (0)1865-794449

THE ECOLOGY OF EXPANSION
AND ABANDONMENT
Medieval and post-medieval agriculture and
settlement in a landscape perspective

Graphic Design
Thomas Hansson
Maps
Henrik Pihl
Cover photo
Sven Waldemarsson
Print
Grahns tryckeri AB, Lund, Sweden 2007
National Land Survey maps
© *Lantmäteriverket, S-801 82 Gävle, Sweden. Dnr l 1999/3*

© Riksantikvarieämbetet 1:1

ISBN 978-91-7209-441-3

PREFACE

Although this is a single-author book it is the fruit of collaboration. Many colleagues have been involved, from the initial archaeological excavations and peat coring, via pollen counting and data processing, to seminars and discussions. In particular I want to thank Mats Anglert, Björn E. Berglund, and Janken Myrdal for stimulating discussions and inspiration. I also want to thank Leif Björkman, Anna Broström, Pär Connelid, Anna Dahlström, Stefan Larsson, Sten Skansjö, and Bo Strömberg for support and discussion. Leif Björkman and Nils-Olof Svensson are gratefully acknowledged for their high-quality pollen-analytical work and Marie-José Gaillard for giving me access to unpublished pollen data.

The research has been carried out within the frame of commissioned archaeology and has been financed by the Swedish National Road Administration.

Per Lagerås
Lund, 2006

CONTENTS

INTRODUCTION

. . . .

The cultural landscape is a complex and fascinating reflection of society. With all its characteristics and peculiarities the landscape tells the story of its inhabitants, their activities, decisions, and ambitions, and, of course, it also to some degree limits them. This interesting relationship between humans and their environment is the basic factor that enables us to use studies of landscape change as a tool, not only for the study of landscape change as such, but also for the study of human history and society.

In this book I discuss landscape change in southern Sweden during the last millennium. It will be evident from the discussion that the landscape has changed in character several times and that some of these changes have been rather dramatic. Throughout the book I will interpret and discuss these specific landscape changes – their character, intensity, and timing – and I will also on a more general level discuss long-term trends and periodicity. The discussion will provide a long-term perspective to our understanding of the present landscape and may hopefully also be useful as a historical background for discussions on nature conservation issues.

One major result presented in the book is that there has not been a one-way development towards a gradually more open landscape, which might be expected, but rather an interplay between periods of deforestation and agricultural expansion on the one hand and periods of decline and reforestation on the other. Based on interpretations of new data from detailed case-studies, but also on the re-examination of earlier published data from a large number of sites, I will argue that there has been major abandonment of settlements and agriculture in particular during two distinct periods, one during the 14th and

15th centuries following the Black Death of the late Middle Ages, and one during the late 19th and the 20th centuries in connection with the depopulation of the countryside and the introduction of modern silviculture.

In order not to limit the discussion to a one-dimensional description of landscape change through time, I attempt to understand the underlying causes and processes. Empirically the book is based mainly on the results of palaeoecology – in particular pollen analysis – but important additional information has also been gained from several other disciplines, such as archaeology and history, and the discussion presented clearly has an interdisciplinary character. With this approach I use the study of landscape change to discuss historical changes in society, from the early-medieval expansion until today. A methodological and conceptual starting point has been that to understand underlying causes and processes, each specific landscape change has to be studied in its own historical context. This means for instance that two recorded landscape changes during two different periods, which at first glance may look similar, cannot a priori be interpreted in the same way, simply because the underlying society has been different. In other words, any historical landscape change has to be studied and interpreted in the light of what we know about society during that particular period. Otherwise, the interpretations will be too vague and much of the causal relationship between landscape change and societal change will remain hidden to us. This approach has been used before in palaeoecological and other landscape-historical studies, but perhaps not so explicitly and thoroughly for the last millennium as in this book.

As mentioned above, much of the discussion in the book is based on palaeoecology. It is a well-established discipline that focuses on long-term landscape change, vegetation, fauna, and land-use development (for an introduction to Quaternary palaeoecology, see for example Birks & Birks 1980). Palaeoecology with focus on the Holocene has strong links with other disciplines with a similar time-perspective, such as palaeoclimatology, geology, archaeology, and history, and also, of course, with botany, zoology, and with what may be called neo-ecology, i.e. the ecology of the present. In Sweden, several of these links have become stronger during the last few decades, due to successful interdisciplinary projects and collaborations (e.g. Berglund 1991, Lagerås 2000, Berglund & Börjesson 2002, Emanuelsson et al. 2003).

A crucial point for such collaboration and for landscape-historical studies in general, is the selection of scale and resolution, both temporal and spatial. Linking palaeoecology to prehistoric archaeology is relatively easy, at least in this respect, as the two disciplines work with similar scales. The linking to history and to archaeology of the historical period is more difficult (the historical period in Sweden is approximately the last one thousand years; see figure 1). One reason for this is that the temporal scale of most palaeoecological studies is too coarse, i.e. the time-resolution is too poor and the absolute chronologies, which are based on radiocarbon dates, are too uncertain for comparison with the rather detailed historical record. Furthermore, in the cultural landscape as well as in society as a whole, both the rate and the frequency of change have increased through time, so that they are much higher during the historical period than during most of prehistory. The majority of pollen analyses and other palaeoecological studies that cover several thousands of years, sometimes the entire Holocene, are therefore not detailed enough to provide any meaningful contribution to discussions on the more recent past. Also the link from palaeoecology to neo-ecology, and in particular to discussions on nature conservation issues in the present landscape, are rather weak, basically for the same reasons. During the last fifteen years or so, the situation has improved, with a number of pollen-analytical studies of stratigraphies from small ponds and peatlands with a focus on local vegetation history (e.g. Bradshaw & Hannon 1992, Björkman 1996a, Lindbladh 1998). There are still, however, very few examples where specific questions about changes in society and landscape in the last millennium have been addressed by palaeoecological studies.

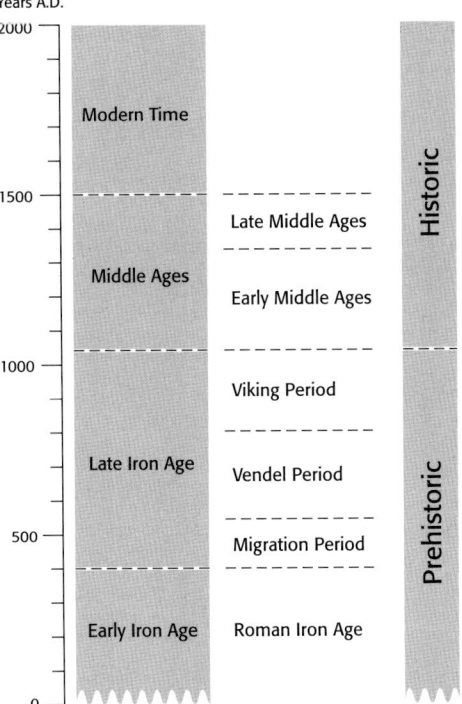

Fig. 1. *The Swedish time scale for the last 2000 years.*

This book deals with the landscape development of an upland area in southern Sweden which was not colonized until the medieval expansion, but which has undergone several changes since then. In the book I present a rather detailed picture of vegetation, land-use, and society in this area, with a focus on the dynamics of agrarian expansion and regression, farm establishment and abandonment. Although local in their empirical basis, many of the phenomena that are discussed – such as medieval colonization and expansion, periods of regression and abandonment, landscape effects of iron production and coaling, the Little Ice Age, heathland formation, the introduction of silviculture – are relevant to large areas of northern Europe and beyond.

The investigation area is situated in the northern part of the province of Scania and is to a large degree covered by coniferous forest. It is an upland area with a rather hilly topography, and with altitudes ranging between 80 and 160 m above sea level. The Quaternary deposits are dominated by sandy till, reflecting the hard underlying gneissic bedrock, together with sandy glaciofluvial deposits in the valleys. The most common soils in the area are podsols and unstable brown earth. The area is also rich in peatlands, in particular bogs, usually with sparse pine (*Pinus sylvestris*) vegetation.

Today, the bogs and the podsols, and the fact that forestry is far more important than crop cultivation, are reflections of the area's rather poor climatic conditions. In a south-Swedish perspective the area is relatively cold. The mean year temperature is 6.0–6.5°C (January: -2.0°; July: 15°), which may be compared to 7.5–8.0°C in the agricultural districts of southern Scania. The climate is not only cold but also humid and cloudy. With a mean annual precipitation of 900 mm the area has the highest precipitation in Scania (the agriculture districts have approx. 500–650 mm per year). The vegetation period is calculated to be 197 days, which may be compared to 219 days in the climatically most favoured parts of Scania (these data are based on instrumental data for the period 1961–1990; Blennow et al. 1999). To summarize this climatic information, the investigation area may simply be referred to as a marginal area, at least in the perspective of modern agriculture.

Thanks to the richness of well-preserved peatlands, the area has very good conditions for palaeoecological studies. The empirical backbone of the book is a series of four local pollen diagrams from small peatlands (figure 2). The pollen-analytical studies were part of a large rescue archaeology project that was run by the National Heritage Board, Archaeological Excavations Department, in Lund (Riksantikvarieämbetet, UV Syd) and occasioned by the rebuilding of the E4 motorway. The pollen-analytical and other palaeoecological investigations that are presented in the book, were planned and performed in close collaboration with archaeologists (and also with historians

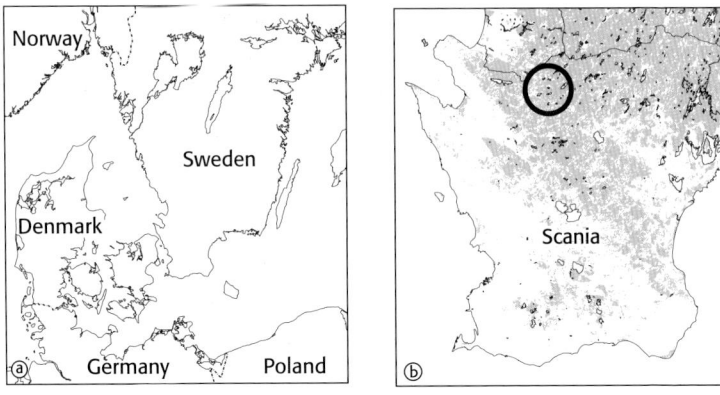

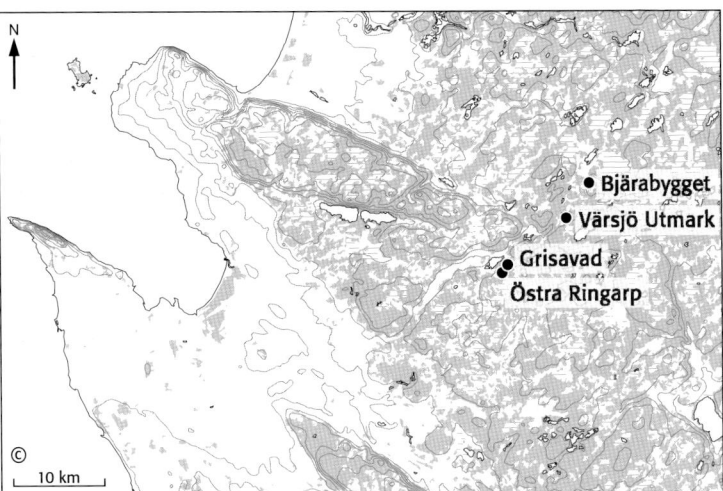

Fig. 2. Maps showing (a) southern Sweden and neighbouring countries, (b) the province of Scania, and (c) the investigation area in the uplands of northern Scania. Grey shading in (b) and (c) indicates present-day forest cover, and hatching in (c) indicates peatlands. The coring sites for pollen analysis are indicated in (c).

and social geographers), who were working in the same area, with the same time perspective, and, partly, with the same questions. The selection of sites for pollen analysis was based on extensive reconnaissance coring in the peatlands, and on judgements based on the distribution, location, and character of archaeological features, as well as on the scientific outcome of excavations (Lagerås 2003a). Many of the archaeological features that were investigated were remains of ancient land-use – such as clearance cairns, farm buildings, iron furnaces, slag heaps, charcoal production sites, tar pits, etc – which provide important information on past landscapes.

The coring sites chosen for pollen analysis were situated in close proximity to excavation sites. (The different coring sites are presented in more detail in Appendix 2.) The aim was to capture pollen from past vegetation and land-use in the documented and excavated areas in order to enable a direct comparison between archaeological and palaeoecological data. The excavations contributed with landscape-ecological information in different ways, for example by dating and interpretation of clearance cairns or iron production sites. A specific and very useful contribution came from the large number of macroscopic charcoal pieces that were collected from different features and contexts, and which were identified to tree species or genera and also were radiocarbon-dated. These high quality data provide direct information about fire clearances and they are an important complement to the pollen diagrams (the method was presented by Lagerås & Bartholin 2003).

The pollen source area of any pollen diagram is dependent on basin size, i.e. the size of the sampled lake or peatland (e.g. Jacobson & Bradshaw 1981, Prentice 1988, Sugita 1994). Although complicated and non-linear, this relationship is direct in the sense that large basins have large pollen source areas and small basins have smaller ones. However, when it comes to peatlands, basin size may be important for the influx of terrestrial pollen but even more important is the distance between the coring point and the edge. Small peat hollows have a great potential for studies of local vegetation changes, which has been emphasized by many authors (e.g. Bradshaw 1988, Björkman 1996a, Lindbladh 1998), but also larger peatlands with edge core location may be used for such a local approach (Edwards 1991). The key point is the distance between the coring point and the edge. By comparing pollen diagrams from two cores from the same peatland, Lagerås et al. (1995) showed how a strong pollen influx of cultural indicators in a core fifty metres from the edge was almost non-existent in another core taken fifty metres further out.

A problem with edge core location is that it may be difficult to find long stratigraphies. Peat sequences are usually thicker – and hence the period that may be studied by pollen analysis longer – in the central parts of peatlands, which is why the vast majority of pollen diagrams is based on central cores. Studies of the vegetation development of the entire Holocene in particular are dependent on such long stratigraphies. However, since the studies presented here only deal with the last two millennia and have a strong focus on the last millennium, it was possible to find suitable stratigraphies close to the edge of the sampled peatlands. This shows the importance of a connection between scientific aim and sampling strategy, and it helps to explain why most pollen diagrams of long-term vegetation development give too few details to be useful in discussions on later periods.

The peatlands in the investigation area are in general well preserved, but they are not completely undisturbed by modern activity. Ditches dug through the peatlands or close to them, in most cases to enable silviculture, have resulted in draining and lowering of the water table. The result of this drainage is increased humification and compaction of the uppermost peat. Analyses show, however, that the pollen record is not destructed – it is compacted but complete. By very dense subsampling it has therefore been possible to produce pollen diagrams with a high temporal resolution even for the upper parts of the peat stratigraphies.

Throughout the book, focus will be on the dynamics of farm establishment and agrarian expansion on the one hand

and regression and abandonment on the other, and the land-scape diversity, not only in space but also through time, will be emphasized. Earlier pollen-analytical studies have often failed to reveal this diversity due to their regional character and/or poor temporal resolution. Regional pollen diagrams are suitable for studies of long-term change and vegetation trends (e.g. Berglund et al. 1996a), but in a more short-term and local perspective they give a too static picture of the landscape. On the other hand, also very local pollen diagrams based on sampling of small hollows, known as close-canopy sites (Bradshaw 1988), may be problematic to interpret (in the way that, for instance, a single oak tree which grew close to the sampling point and lived for 300 years may be misinterpreted as a 300-year period of oak woodlands). They are suitable for stand-scale studies, but may be too local for other research aims. Pollen diagrams with a character somewhere between regional and local, i.e. with a relevant pollen source area of a few hundred metres, are probably the most appropriate for studies of farm establishment and abandonment, and therefore the diagrams presented in this book are of this kind. Such pollen diagrams are sometimes referred to as extra-local (Jacobson & Bradshaw 1981), but for simplicity I use the term local throughout the book. Working at farm scale in palaeoecology is fruitful as much of the landscape diversity depends on decisions and efforts made at farm level.

Before ending this introduction, some words on the structure of the book may be useful. The main part is a series of chapters that focuses on different periods and phenomena in a more or less chronological order. The chapters are *Medieval Colonization and Expansion, Late-medieval Decline, 16th Century Re-expansion, Iron and Charcoal Production, Man-made Heathland: Birth and Decline, 19th Century Crofts,* and *The Return of the Forest.* After these comes *Synthesis* in which I summarize the results and present some general models and conclusions about landscape diversity and about the process of expansion and abandonment. After the references there are three appendices. In *Appendix 1* all the methods used are presented, while in

Appendix 2 total pollen diagrams with all identified taxa are presented site by site, together with site-specific primary interpretations. *Appendix 3* is a list of English plant names with their Latin and Swedish equivalents.

Throughout the book, discussions and arguments in the text are accompanied by a number of special pollen diagrams. They are based on the same data as the total pollen diagrams in Appendix 2, but to be clear and easy to grasp, they only present a few selected graphs relevant to the discussion. To enable comparison, all the pollen diagrams cover the same period, i.e. the last two thousand years, even though the original analyses reached further back in time. The combination of different graphs in special diagrams has not only been a matter of presentation, but it has also been an important tool in the interpretation process.

MEDIEVAL COLONIZATION
AND EXPANSION

·········

Before colonization:
Prehistoric herding and transhumance

The investigation area is one of very few areas in southern Sweden that totally lack prehistoric monuments. Most other areas show a variety of more or less monumental graves – such as burial mounds, stone settings, burial cairns, stone cists, megaliths, etc. – in particular from the Iron Age and the Bronze Age, but in some areas also from the Neolithic. Based on results from the National Survey of Ancient Monuments, the archaeologist Åke Hyenstrand showed that there is a monument-free zone along the southern border of the South-Swedish Uplands, stretching more or less along the border between the province of Småland in the north and the provinces of Halland, Scania and Blekinge in the west and south (Hyenstrand 1979). Before A.D. 1658 this border between the provinces was the national border between Denmark and Sweden, which in combination with the absence of prehistoric monuments gives the impression of an uninhabited borderland with a very long continuity.

The maps of prehistoric settlement patterns presented by Hyenstrand were based on monumental graves and other traditional archaeological features. Since then, a new category of prehistoric remains has been discovered, namely clearance cairns. These cairns are ancient heaps of stones cleared from arable fields sometime in the past. They are typically 2–6 m in diameter, 0.2–0.5 m high and covered by mosses or moor litter, and they are very common in the forest regions of southern Sweden. Of course, these clearance cairns had been noted before, and even in early literature they were regarded as remains of ancient agriculture, for instance by Carl von Linné (Linnaeus 1751), but it was not until the 1980s that archaeologists realized that they may be of prehistoric origin (e.g. Gren 1989). This important conclusion resulted in increased attention, and during the following decade several areas with clearance cairns were investigated (e.g. Jönsson et al. 1992, Connelid et al. 1993, Lagerås 2000). From these studies we now know that many clearance cairn areas indeed are prehistoric, in particular from the early Iron Age, and also that some areas are medieval or even younger. Sometimes there are prehistoric graves within or in close connection to the clearance cairn areas, but sometimes there are not, even when dating has proved that they are of prehistoric origin. The distribution of clearance cairn areas may therefore be an important complement to the monumental graves, and they may fill in empty areas on maps of prehistoric settlement patterns.

Before the project presented in this book started, large areas of clearance cairns had recently been investigated some kilometres to the south, on the rim of the upland in the parish of Rya (Lagerås et al. 2000). Dating showed that these areas reflect an agrarian expansion during the Roman Iron Age (c. A.D. 200) and that they were in use for some hundred years until the end of the Viking Period (c. A.D. 1000). Excavations among the clearance cairns also revealed Iron Age long houses, hearths, slag, etc., but no definite graves. Similar results were obtained in Hamneda, close to the river Lagan in the province of Småland, where large areas with clearance cairns and settlement remains, but no graves, could be dated and connected to a Roman Iron Age expansion (Lagerås 2000, Lagerås & Bartholin 2003).

Also in the present investigation we found clearance cairns but with a somewhat different appearance. The cairns themselves looked similar to the ones in Rya and in Hamneda but they made up smaller areas. Initially in the project, we thought that they might reflect small-scale or short-lasting Roman Iron Age cultivation, and that they were to be interpreted as outposts of the agrarian expansion that had resulted in the larger clearance cairn areas to the south. Now, with the results of excavations and dating at hand, we can say that this initial hypothesis has turned out to be wrong. The clearance cairns of the investigation area reflect a later agrarian expansion, mainly during the Middle Ages, but in some places starting as early as the Viking Period or somewhat earlier, that is during the last centuries of the first millennium A.D. (figure 3). The Roman Iron Age expansion that resulted in cultivation and settlement at Rya was obviously limited to the border zone or the "first step" up in the upland, and did not reach the investigation area. It was not until the medieval expansion (in a broad sense) that this more remote upland area witnessed cultivation and permanent settlement. In this respect, the picture given by Hyenstrand still stands – the monument-free zone does really reflect a zone with no, or at least very few, prehistoric settlements.

So what about the period before the medieval expansion? As mentioned above, no traces of cultivation or permanent settlements have been found, in spite of careful archaeological surveying and a very large number of excavations. But there are traces of humans. First of all, quite a large number of Mesolithic sites have been found and investigated within the project, but as they are from before the introduction of agriculture they are not dealt with here (see B. Knarrström 2007). Apart from these Mesolithic remains, a relatively large number of hearths have been found. A compilation of radiocarbon dates from these hearths shows that they represent a long period, lasting from the early Bronze Age (c. 1400 B.C.) almost to the present (figure 4). Very few finds, such as pottery or flint, have been made in connection to them and their age would have been impossible to judge without radiocarbon dating. The question is what these anonymous hearths represent. Humans have obviously been visiting the area for a long time, but for what purpose? Possible explanations are that they were just passing by, crossing this upland area on their route between for example the plain areas of Scania and Halland, or, more likely, that they used these upland areas as hunting grounds. However, the pollen diagrams add some important information.

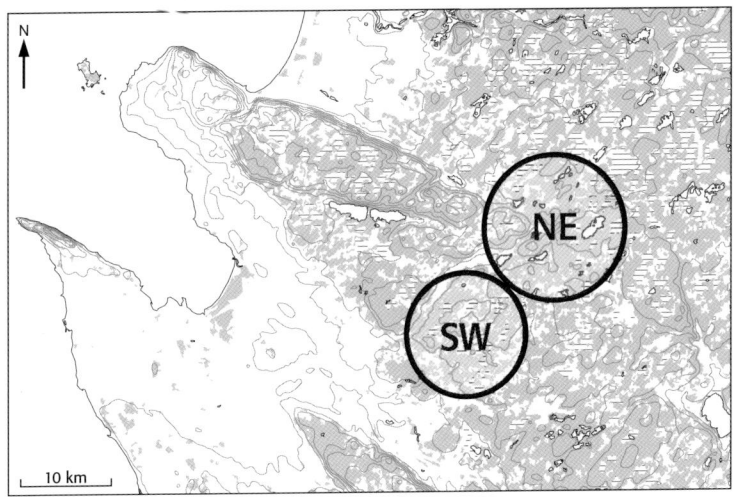

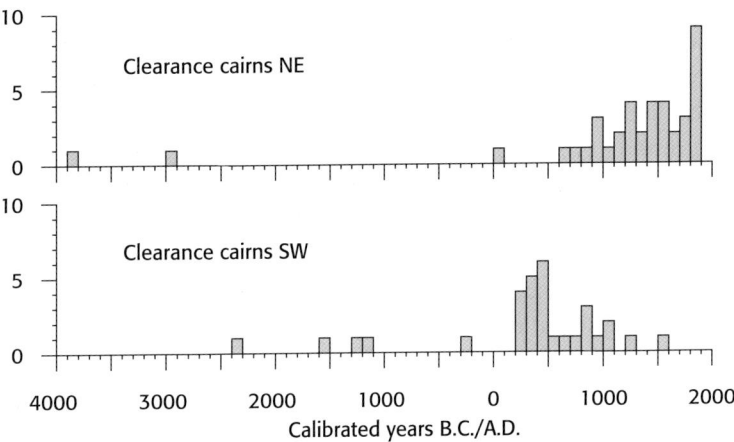

Fig. 3. *Compilation of radiocarbon dates from clearance cairns investigated within the present project (NE), and from clearance cairns investigated within an earlier project at lower elevation (SW) (Lagerås et al. 2000). The compilation is based on radiocarbon-dated macroscopic charcoal collected from the bottom layers of the clearance cairns. Such charcoal may originate in clearings that were carried out when the ground was prepared for cultivation, but it may also originate in earlier fires that had no connection with cultivation or stone clearance. The latter is probably true for the scattered early dates. The bars show the number of dates within each 100-year interval, and are based on the mid-points of the calibrated one-sigma intervals. The two bar charts are based on 40 and 31 radiocarbon dates, respectively.*

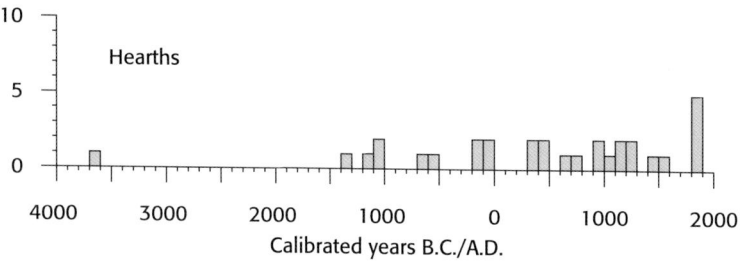

Fig. 4. *Compilation of radiocarbon dates from hearths investigated within the project. The bars show the number of dates within each 100-year interval, and are based on the mid-points of the calibrated one-sigma intervals. The bar chart is based on 31 radiocarbon dates.*

All of the four local pollen diagrams from the area show a weak but significant grazing signal long before the medieval expansion. The diagrams presented throughout this book cover the last two thousand years and they show a more or less continuous curve for *Plantago lanceolata* (ribwort plantain), which is considered to be a reliable indicator of grazing and possibly hay mowing (e.g. Behre 1981, Gaillard et al. 1992, Hjelle 1999), from the bottom of the diagrams. At the Östra Ringarp site, however, the original analysis reaches much further back in time than the others, and it is the only site where the diagram captures the introduction of grazing (figure 5). The diagram shows some pollen grains of *P. lanceolata* from between 4000 and 3000 B.C., that is from the Early Neolithic, while a continuous curve begins at c. 1700 B.C., that is the Early Bronze Age. From about the same levels there is also a more or less continuous curve for *Rumex acetosa/acetosella* (sorrel). To sum up, we may therefore conclude that we have local evidence of grazing from c. 1700 B.C. onwards, a date which is relatively close to the first appearance of hearths.

The occurrence of *Plantago lanceolata* pollen and some other herb pollen taxa from c. 1700 B.C. onwards clearly indicates grazing in the area, but the grazing pressure must have been very low as the high frequencies of tree pollen indicate a

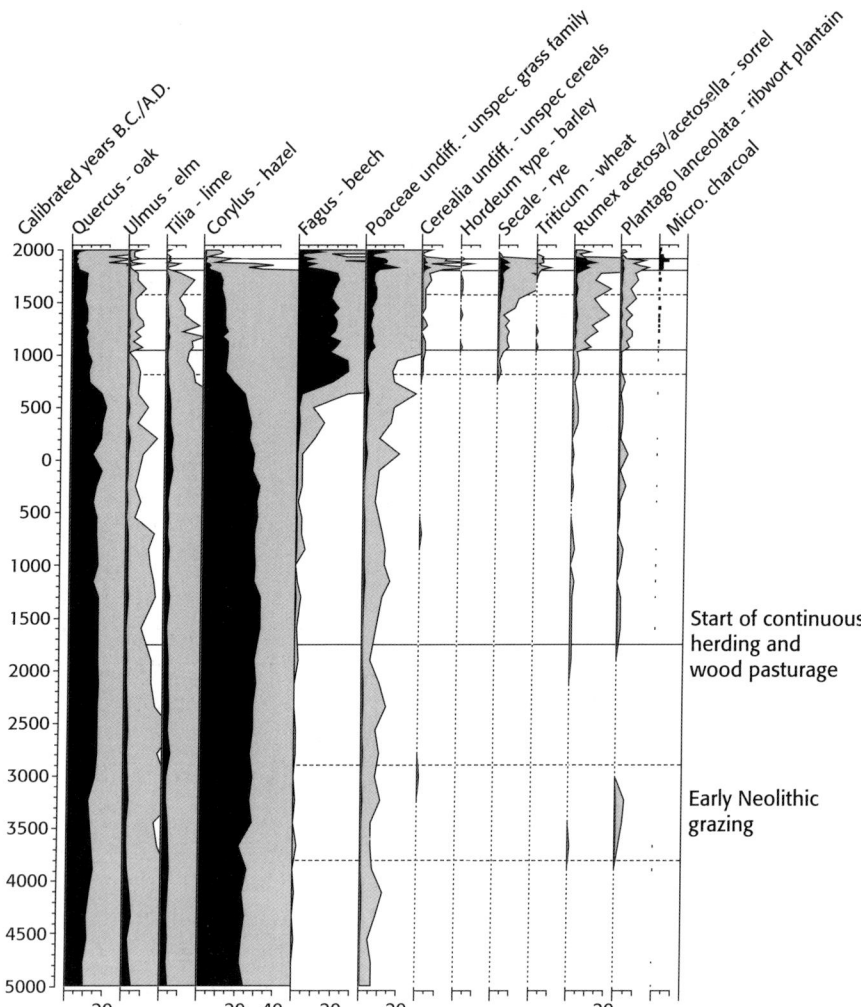

Fig. 5. Pollen diagram from Östra Ringarp with selected taxa. This diagram as well as other pollen diagrams throughout the book is based on the total pollen diagrams with all identified taxa, which are presented in Appendix 2. The graphs show pollen percentages (black) and ten times exaggeration (shaded), and they are based on the total pollen sum. While the other pollen diagrams throughout the book cover only the last 2000 years, the diagram in this figure reaches much further back in time to capture the introduction of herding. In addition to pollen percentages, the diagram also shows the amount of microscopic charcoal in the pollen samples (% of pollen sum).

landscape almost completely covered by forest. The tree species composition of these forests was dominated by oak (*Quercus*), birch (*Betula*), lime (*Tilia*; this taxon is insect-pollinated and thus underrepresented in pollen diagrams), hazel (*Corylus*), and on damp soil alder (*Alnus*). It is important to note, however, that the degree of openness tends to be underestimated in percentage pollen diagrams during periods dominated by forest (Broström et al. 1998). The reason for this is that the influx of not only herb pollen but also tree pollen responds positively to an opening-up of the dense forest, while in more open landscapes the influx of tree pollen responds negatively to further deforestation (Aaby 1994). Nevertheless, we may conclude that the area was mainly forested and that grazing had probably only resulted in minor openings and thinning out of the tree cover, even if we are not able come up with a more precise quantification.

Most likely this grazing during the Bronze Age and the Iron Age was in the form of herding. Weak occurrences of grazing indicators in pollen diagrams, not least from the South-Swedish Uplands, have also been interpreted as transhumance or other forms of herding in earlier studies (Almquist-Jacobson 1994, Lagerås 1996a). But those interpretations were based almost entirely on pollen data. In the investigation area, however, we now have a strong combination of data: (1) a weak but continuous grazing signal in local pollen diagrams from the Early Bronze Age and onwards; (2) a more or less continuous series of hearths that begins in the Early Bronze Age; (3) a total lack of prehistoric grave monuments; (4) a lack of houses or other settlement remains in spite of numerous excavations; (5) a lack of clearance cairns dated to this period even though they are numerous from later periods, and (6) no or very little cereal pollen. All together these data and observations constitute a good basis for an interpretation of herding. Shepherds from more low-lying areas with permanent settlement probably brought their cattle up to these uplands for grazing in the summers, and the numerous but anonymous hearths are remains of their simple campsites. Also, in a recently published study in the province of Östergötland, hearths were interpreted in a similar way (Petersson 2006). In northern Scania it is justified to classify this system as transhumance, which in contrast to nomadism implies that the same grazing grounds were visited again and again on a seasonal basis.

As a background to the next section, the important thing to note here is that the area was not complete wilderness before the medieval expansion. Even though there were no or very few settlements, people and their livestock had been using it for more than one thousand years. They knew the area, its character, resources and pathways, and they had probably socialized it with place names and stories.

Medieval expansion in a North-European perspective

In the Middle Ages the upland area in northern Scania was subject to major changes. Permanent settlement and agriculture were established, large-scale iron production started, churches and castles were built. Even though the area had been used for seasonal herding for two thousand years, and in that sense was no wilderness, the medieval expansion was so thorough and of such high impact that it is justified to call it colonization. Interestingly enough, this change in local society and landscape was not an isolated phenomenon, but rather the local expression of a general expansion that swept through Europe at this time. Before we look at the colonization of the investigation area in detail it may therefore be fruitful to present a broader perspective.

The background or starting point for the medieval expansion was the decrease in population that Europe witnessed after the fall of the Roman Empire. There may have been several factors behind this decline, but the most important one was probably the plague epidemics which haunted the European population during the 6th century. Starting with the first major outbreak in 543, its deadly waves swept back and forth over the Continent for several decades. It has not been proved if this epidemic reached Sweden or not, mainly due to the lack of such early written sources, and different opinions have been put forward. We can be sure, however, that the 6th century was a period of decline also here, reflected for example in farm abandonment and reforestation in marginal areas, which is evident from pollen diagrams (Lagerås 1996).

After the decline a long period of expansion and population growth followed in large parts of Europe. It started in the 7th century and lasted for almost seven hundred years, until the next large-scale plague epidemic – the Black Death of the 14th century (see the chapter *Late-Medieval Decline*). The expansion as a whole is here referred to as the medieval expansion, although the chronological nomenclature differs from country to country. In Continental Europe the Middle Ages is usually defined as beginning as early as the 7th century or even earlier, while in Sweden the same period is divided into the Vendel Period (550–800), the Viking Period (800–1050), and the Middle Ages (1050–1500) (figure 1). A similar time-scale is used in Denmark. But differences in nomenclature and time-schemes are of minor interest. More important is the fact that the medieval expansion affected and included not only Central Europe but also, for example, Denmark and Sweden.

In an attempt to give a general outline of the expansion in a European perspective, the period c. 600–800 may be distinguished as an introductory phase. It was characterized by internal expansion and population growth in central areas and possibly also external expansion into marginal areas (Duby 1973). To a large degree this was caused by a population rise and expansion which filled up an empty space left by the decline of the 6th century, and which was not accompanied by any substantial agro-technological development. After a short stagnation the expansion began again with accelerating population growth, major agrarian expansion, and colonization of most hitherto uninhabited forest areas. In this very expansive phase, which lasted for several hundred years but seems to have been most pronounced around the 12th century, society underwent a series of gradual but fundamental changes. Many of these changes were related to each other but their causal relationships are difficult to sort out. Among the most important changes and processes were urbanization, the introduction of a monetary system, the spread and acceptance of Christianity, and the establishment of a feudal system. A process that may have been particularly important for the agrarian and settlement expansion was the replacement of thraldom (i.e. slavery) by a system based on tenant farmers (Sw: *landbor*) (Lindkvist 2003). This was a complex process, which was slow and gradual with intermediate stages during the Early Middle Ages (Myrdal & Tollin 2003). As tenant farmers the peasants were still dependent, but after they had fulfilled their duties in day-labour and paying taxes they had the possibility of improving their own standard through hard work. This was an important difference from thraldom. As a consequence, forests were cleared for agriculture, production increased, and population numbers rose. The last thralls in Sweden were released during the early 14th century, but in most parts of Sweden as well as Denmark thraldom was in strong decline as early as the 11th and 12th centuries.

The rise in population was vigorous, in particular during a period centred on the 12th century, but it is difficult to quantify in absolute numbers. In England it has been estimated that the population more than tripled from the end of the 11th century (based on the Domesday Book of 1086) to the mid-14th century (Duby 1973), and other parts of northern Europe may have witnessed a similarly vigorous rise. In Denmark and Sweden we do not have any written sources that can be used for population estimates of that time, but a vigorous rise is indicated, for example, by the fact that the number of villages was at least doubled during the same time (Myrdal 1999a). In southern Scania and other central settlement areas the number of grave fields shows that a pronounced rise in population started as early as the 8th century, but most marginal upland areas were probably not affected by the expansion until the 11th and 12th centuries.

This population rise can be seen as an important factor behind the agrarian expansion. On the other hand, population rise would not have been possible without a development of agricultural tools, implements and techniques. The improvements within the agricultural system at this time laid the foundation for a larger population and also provided a surplus that could be invested in cities, castles and churches. Technical developments within agriculture may to a large degree be explained by a greater availability of iron, with which stronger, sharper and more efficient implements and tools could be made. Most important were probably the spade with an iron tip and the ard with a longer share, both used for tilling (Myrdal 1997). Also the plough was introduced, in Scania as early as the 11th century, but the use of it was for long confined to fertile plains. In upland areas, where the cultivated fields were rich in stones and boulders, the spade and the ard were dominant and were still in use all the way up to the beginning of the 20th century. Another important improvement was the development of a scythe with a long blade, which made the collection of hay-fodder for winter-stalled livestock much more efficient. Finally, the introduction of larger and heavier iron axes aided deforestation and the transformation of woodlands to fields, pastures and meadows.

In addition to the technical improvements and the changes in society which have been discussed so far, a factor that may also have contributed in a positive way to the medieval expansion was climate. Attempts to ascribe societal changes to climatic events have a long tradition in archaeology and history (e.g. Wendland & Bryson 1974), but the interpretations of causal relationships have often, for good reason, been criticized for being too simple. Even if most of us agree that climate sets important frames to human activity, opinions differ to what degree climatic changes really have been decisive for development in society. However, in climatically marginal areas, where summer temperatures and the length of the growing season are limiting factors for agriculture, changes in climate have certainly played an important role and may help to explain periods of establishment on the one hand and abandonment on the other.

Earlier studies based their comparisons on rather poor climatic data, but great advances have been made within palaeoclimatic research during the last few decades. The amount of proxy data is increasing all the time – for instance from inland

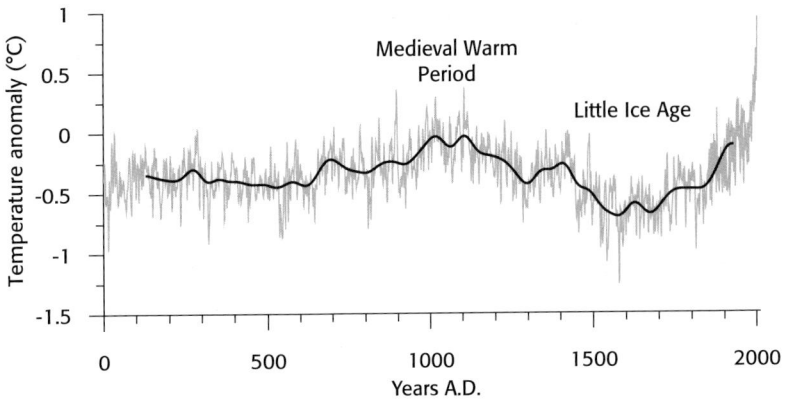

Fig. 6. Mean temperature curve for the last 2000 years in the Northern Hemisphere presented as anomalies with respect to the 1961–1990 average (Moberg et al. 2005). The curve is based on high-resolution tree-ring data, low-resolution proxies from lake and ocean sediments, and for the last two centuries instrumental data.

ice cores, deep sea sediments, and tree rings – and, what is equally important, progress has been made in how to combine data of different kinds. The best temperature curve for the Northern Hemisphere available at the moment is probably the one published by Moberg et al. (2005), which is based on a combination of high-resolution tree-ring data and low-resolution proxies from lake and ocean sediments (figure 6). As evident from this curve, the period 600–1100 was characterized by a gradual and step-wise increase in mean temperature, with two pronounced warm peaks around 1000 and 1100. After the temperature optimum at 1100, a cooling trend started which use to be referred to as the Little Ice Age, and which finally reached bottom values in the 16th and 17th centuries. Since then there has once again been a gradual increase in temperatures, and at the right end of the diagram the curve rises steeply, indicating anthropogenic warming due to the so-called greenhouse effect. What is most interesting for the discussion here is that the medieval expansion obviously coincided with a period of increasing mean temperatures. It is also notable that the period c. 950–1150 was warmer than any other period of the last two thousand years, with the exception of the greenhouse temperatures of the last century. The Norse colonization of Greenland in the late 10th century, with the establishment of settlement and farming which lasted until the early 15th century, is the most striking reflection of this Medieval Warm Period. However, also in climatically less extreme environments, such as the South-Swedish Uplands, the medieval expansion process was probably supported or reinforced by the relatively favourable climate of that time.

Colonization and farm establishment in northern Scania

From the general background of climate and medieval expansion in a European perspective presented above, we may now take a closer look at the investigation area in northern Scania and explore local evidence of colonization. To begin with, there are no written sources from the area that reach far enough back in time to cover the colonization process. The earliest preserved document with a local connection is a letter by the Danish king Erik Menved, which was written in 1307 at a royal castle in Örkelljunga (Ödman 1995, Skansjö 1997a). Not much is known about the castle except that it was burnt down in 1316, probably only a few decades after it was built. The barely visible remains of it have been identified on a forested hill in Örkelljunga, but no investigation has been carried out. There is also some other mention of the castle in early documents, but it is not until the beginning of the 16th century that we get more detailed information on ordinary settlement. The earliest preserved written documents that give information on local agricultural settlement are two cadastral registers (Sw: *jordeböcker*) from 1514 and 1523 (Skansjö 1997a). For the two parishes Örkelljunga and Fagerhult, which make up most of the investigation area, these documents mention 44 settlement units by name. The size of the two parishes together is 220 square kilometres, which gives an average of one settlement unit per five square kilometres. Most of the settlement units were probably single farms but some may have been double farms or even small hamlets. Thus, these written sources show that there was a well-established settlement in the area by the end of the Middle Ages. They say nothing about how far back in time we may extrapolate this picture, but the existence of a castle indicates that there were some agrarian settlements as early as around 1300. Similar evidence is given by the church in Örkelljunga. It has been modified and rebuilt several times but its oldest parts have been dated to the 14th or possibly 13th century. Since the building of a church during the Middle Ages was dependent on taxes collected from local farmers (Sw: *tionde*), the church is indirect evidence of settlement and agriculture in that early period.

Another source for interpreting the settlement history is the place-names. Some of the names occur in early documents, the earliest one being Örkelljunga (then spelled *Øthknælyung*) mentioned in 1307, but also other names may be tentatively dated based on research and experience from other parts of southern Sweden. In particular the endings are useful for dating purposes. In the area, common place-name endings are *-torp* and *-arp* (meaning 'new farm' or 'moved out farm'), *-hult* and *-alt* (meaning 'forest' or 'grove'), *-bygget* (meaning 'small building in the forest') and *-ljunga* (meaning 'heathland'). These names are regarded as predominantly originating in the Middle Ages or somewhat later (Pamp 1983). Names with the ending *-torp* or *-arp*, may be slightly older, that is from the Viking Period, but most of them are certainly medieval. The oldest types of place-name endings, which are believed to originate in the Early Iron Age or at least before the Viking Period (e.g. *-löv*), are not found in the area. In conclusion, the local place-names indicate settlement establishment mainly during the Viking Period and/or the Middle Ages, but it should be noted that dating based on place-names alone is rather insubstantial.

Turning to the archaeological evidence, it has already been mentioned above that the area was used only for seasonal herding during the Bronze Age and most of the Iron Age, and that

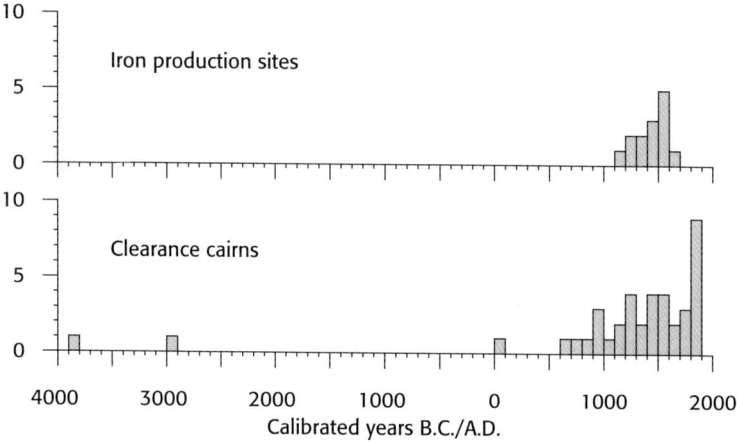

Fig. 7. Compilation of radiocarbon dates from iron production sites and clearance cairns investigated within the project. The bars show the number of dates within each 100-year interval, and they are based on the mid-points of the calibrated one-sigma intervals. The two bar charts are based on 14 and 40 radiocarbon dates, respectively.

it was during the medieval expansion that permanent settlement and cultivation were established. The archaeological results from excavations and surveying within the project shed some light on this colonization process. Most important in this respect are the results of radiocarbon dating from clearance cairns. These cairns are the remains of ancient cultivation in stone-cleared fields, and the charcoal found among the bottommost stones in the cairns is believed to originate mainly from the burning of twigs, trees, and shrubs when a plot was cleared and prepared for cultivation (Lagerås & Bartholin 2003), although some of the charcoal originated in earlier fires. Through radiocarbon analysis of such charcoal it is possible to date when a particular field was cleared and by compiling a large number of dates from several sites in a region it is possible to get an overview of the cultivation history. The compilation presented in figure 7 shows that stone clearance (and hence cultivation) did not start before the medieval expansion, with a continuous series of dates beginning in the 7th century. The geographical distribution of the dates show that the earliest ones, i.e. those from the 7th to the 9th century, are from the south-westernmost part of the investigation area, which has a more central character than the till plateaus in the northeast. These early dates may not necessarily be connected to stone clearance but may reflect fire clearance in connection with temporary cultivation that preceded permanent cultivation and settlement.

Another activity connected to the medieval expansion was iron production. A very large number of slag heaps have been found earlier in the area during surveying, and within the project some sites with iron furnaces and slag heaps were excavated and dated. These dates show that local iron production started in the late 12th century and lasted until the 17th century (figure 7). Also charcoal production started in the Middle Ages but it was more common in post-medieval times. The landscape-ecological effects of iron production and charcoal production are discussed in more detail below in the chapters *Iron and Charcoal Production* and *Man-made Heathland: Birth and Decline*. These activities are mentioned here first of all to show the complexity of the medieval expansion. It is obvious that not only agriculture but also iron production was of economic importance in the area during the Middle Ages, and one of our initial hypotheses was that the suitable conditions for iron production (ore in bogs and lakes, fuel in the forests) were possibly a driving force behind the colonization. According to our dates, however, local production of iron began late in the 12th century, i.e. one or a few hundred years after the first establishment of agriculture and settlement, and a more intense iron production did not start until the 15th century. Probably, the initial colonization of the area was mainly agricultural while a more complex economy was developed later on (this will be further discussed in *Synthesis* below).

From this picture of the medieval colonization and expansion based on archaeological and historical data, we now turn to the pollen diagrams. Four local pollen diagrams produced within the project will be used to interpret the timing and character of the expansion. In order from southwest to northeast, the sites are Östra Ringarp, Grisavad, Värsjö Utmark and Bjärabygget (figure 8; see also Appendix 2 for site descriptions and total pollen diagrams). Together the diagrams give a rather clear picture of the general colonization process of this upland area. At the same time they reveal local differences between the sites.

In general the landscape that met the colonizers in the investigation area was a forested one. How dense the forest was is difficult to say, but it was at least slightly thinned out and it had openings in the canopy due to the seasonal grazing that had been going on for a long time. According to the pollen analyses the dominating tree taxa were oak (probably both common oak *Q. robur* and sessile oak *Q. petraea*), beech (*Fagus sylvatica*), lime (probably mainly small-leaved lime *T. cordata*), birch (both silver birch *B. pendula* and downy birch *B. pubescens*), and hazel (*Corylus avellana*). Less common but still present were, for instance, elm (mainly wych elm *U. glabra*), aspen (*Populus tremula*), and hornbeam (*Carpinus betulus*). On wet ground grew alder (mainly common alder *Alnus glutinosa*), birch, and less frequently pine (Scots pine *Pinus sylvestris*).

The forest composition was rather stable before the colonization but it was not without change. Hazel for example had been gradually decreasing, due to the gradual climatic and edaphic deterioration since the early- and mid-Holocene climatic optimum. (Hazel is technically a shrub but may in an ecological perspective be treated as a tree.) Beech on the other hand was increasing. This tree species expanded relatively late into northern Europe and Sweden and the possible role of climatic and anthropogenic factors behind its migration and establishment have been discussed by many authors (e.g. Björkman 1996a, Küster 1997). The date for the first local expansion of beech in the investigation area varies slightly between the different diagrams, but falls within the time-interval A.D. 400–800. This increase is mirrored in a decrease in oak, which

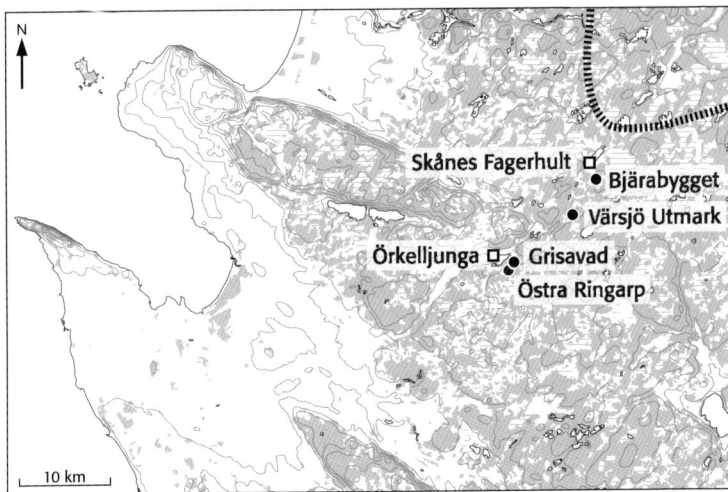

Fig. 8. Map of the investigation area with the four coring sites for pollen analyses. The cities of Örkelljunga and Skånes-Fagerhult are indicated, as well as the medieval border between Denmark and Sweden.

thrive on the same type of soils but obviously lost the competition with the dominant and more shade-tolerant beech.

Looking now at the local expressions of the medieval expansion site by site, we start in the southwest with Östra Ringarp and then move towards the northeast and to gradually higher elevations (figure 8). In the diagram from Östra Ringarp, the medieval expansion starts with a few pollen grains of *Secale* (rye) and Cerealia undiff. (unspec. cereals) at levels dated to the 9th–11th centuries A.D. (i.e. the Viking Period) (figure 9). At the same levels, a peak in SIRM values indicates soil erosion, probably in connection to cultivation. However, the high frequencies of tree pollen indicate a much-forested landscape, so these early cereal pollen grains and the SIRM peak probably reflect only small clearings for temporary cultivation. This initial phase was followed by a major change dated to the 11th century (i.e. the beginning of the Middle Ages according to the Swedish time-scale), during which a more open cultural landscape was established.

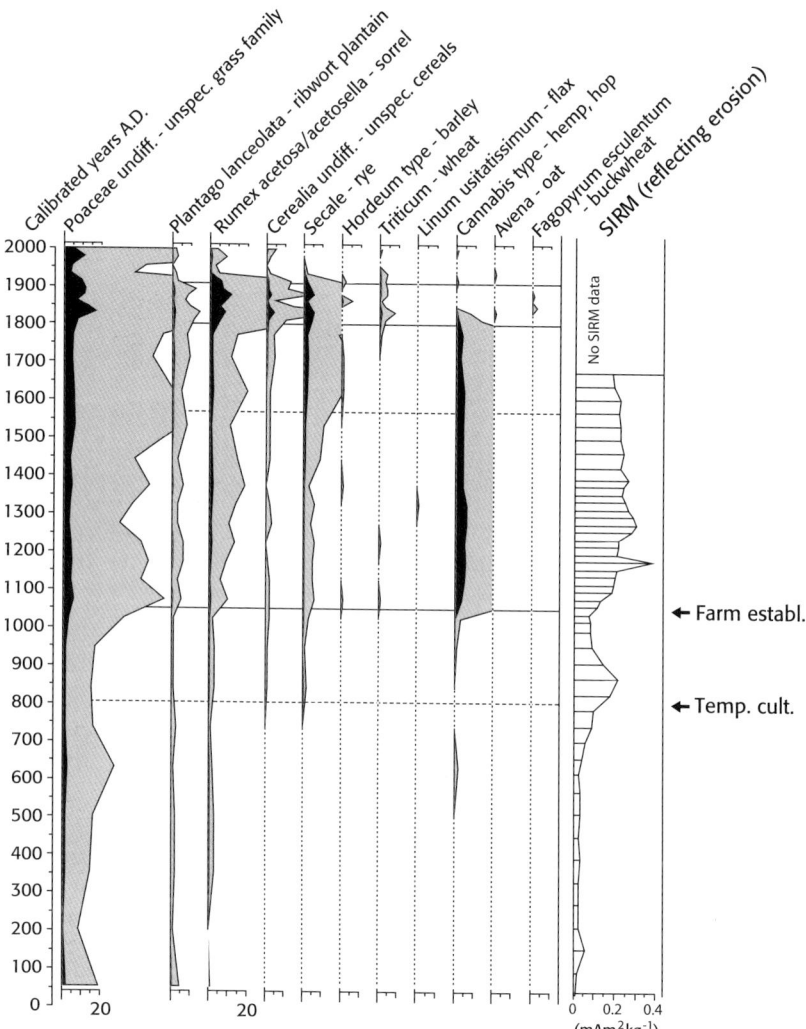

Fig. 9. Pollen diagram from Östra Ringarp with selected taxa. This diagram as well as other similar diagrams throughout the book is based on the total pollen diagrams with all identified taxa, which are presented in Appendix 2. The graphs show pollen percentages (black) and ten times exaggeration (shaded), and they are based on the total pollen sum. The time scale is based on radicarbon dates and linear interpolation. In addition to the pollen graphs, the result of the SIRM measurement of the core is presented, which indicates soil erosion in the catchment area. The levels interpreted as reflecting the introduction of temporary cultivation and the earliest farm establishment are indicated.

From this level onwards the pollen diagram shows strong and continuous curves for several cultivated taxa, and also relatively high pollen percentages of Poaceae undiff. (grass) and some other herbs reflecting grazed or mowed grassland. The changes dated to the 11th century are interpreted as reflecting the establishment of a local farm close to the sampling site. From this level onwards SIRM values remain relatively high, indicating permanent cultivation and settlement. According to the pollen diagram this medieval farm grew rye (*Secale*), barley (*Hordeum*), wheat (*Triticum*), flax (*Linum usitatissimum*), and hemp (*Cannabis* type), which was probably retted in the lake. It also had open or half-open pastures and probably hay meadows characterized by grass and herbs, but with no or very little heather (*Calluna*).

This pollen-analytical interpretation may be supplemented with information from some other sources. An early 19th century map of Östra Ringarp has revealed an ancient strip field pattern close to the sampling site (figure 10) – a type of field pattern that has been discussed in other projects in southern Sweden and which has been interpreted as originating in the Viking Period or the Early Middle Ages (e.g. Connelid 2002). The establishment of a local farm at Östra Ringarp during the 11th century – inferred from the pollen data – can possibly be connected to the establishment of these strip fields. Alternatively, the strip field structure may have been established later, when the population grew and the agricultural land had to be divided among different farmers (see discussion on page 61). According to the map the strip fields were situated northwest and northeast of the sampling site, at a distance of less than 100 m for the closest ones.

Archaeological information from the site is scanty, and except for some test trenches no excavation was carried out within the project. However, back in 1953 a furnace and some other remains of bloomery iron production were excavated in close connection to the sampling site (Wedberg 1981, Englund 1995). By radiocarbon dating the furnace was dated to the 13th or 14th century.

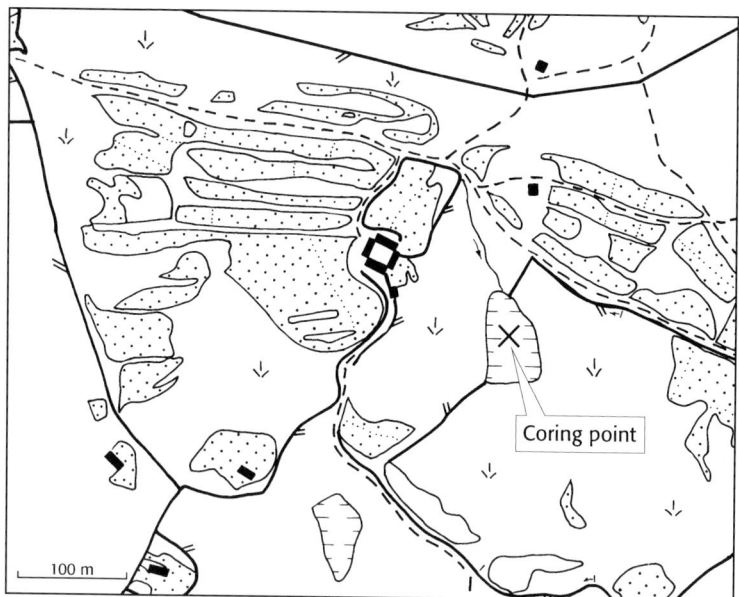

Fig. 10. Map from 1818–1824 of Östra Ringarp showing buildings, arable fields (dotted), and meadows. The coring point for pollen analysis is indicated. Original map redrawn by Pär Connelid.

The local expression of the medieval expansion at Östra Ringarp may thus be interpreted in the following way: A farm was established in the 11th century. There had been no farm on the site before, but the area and its suitability for cultivation was well known due to temporary cultivation that had been practised for some centuries. When the farm was established it grew several different crops and it used grasslands for pasturage and hay production. As a complement to the agricultural economy it was also involved in iron production.

The next site, Grisavad, is situated only 1.6 km northeast of Östra Ringarp. It has a more marginal setting than Östra Ringarp, more or less surrounded by peatland, but the process of medieval expansion has been very similar. The first cereal pollen grains appear at a level dated to about A.D. 800 and at the same time there is an expansion of grassland indicated by an

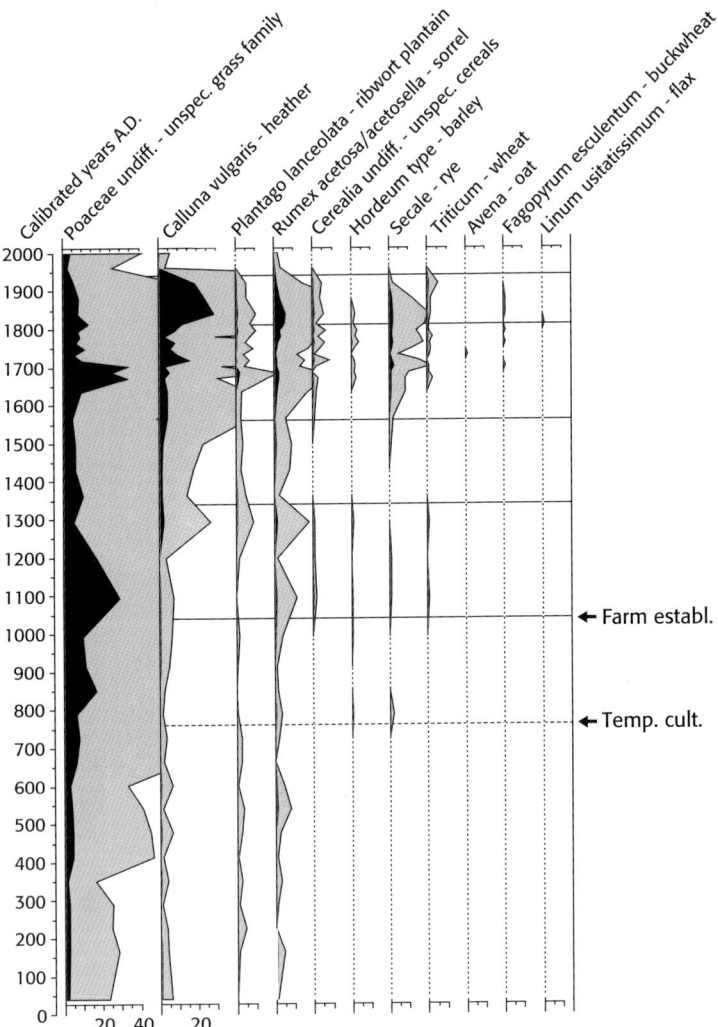

Fig. 11. Pollen diagram from Grisavad with selected taxa. The levels interpreted as reflecting the introduction of temporary cultivation and the earliest farm establishment are indicated. (For details see caption to fig. 9.)

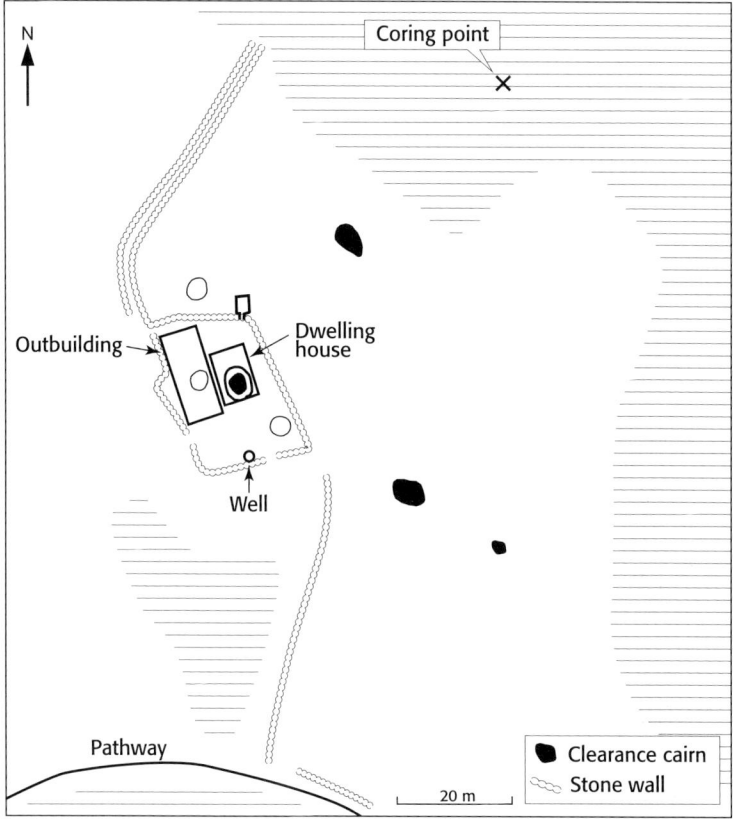

Fig. 12. Map of the Grisavad site with the remains of an abandoned croft. Hatching indicates peatlands. The coring point for pollen analysis is indicated.

increase in Poaceae undiff. (grass) (figure 11). This agricultural expansion may reflect the establishment of a local farm as early as A.D. 800, but the very sparse and discontinuous cereal record rather suggests an initial and temporary cultivation followed by grazing. A continuous cereal record starts at a level dated to the 11th century, where there is also a further increase in Poaceae undiff. Also *Rumex acetosa/acetosella* (sorrel) – which at this time first of all was a weed on arable land – shows a significant increase at the same level. The changes dated to the 11th century are interpreted as reflecting the establishment of a local farm in the vicinity of the sampling site. According to the pollen data this medieval farm grew barley (*Hordeum*), rye (*Secale*), and wheat (*Triticum*), and it had open pastures with field vegetation dominated by grasses. The very high percentages of Poaceae undiff. may indicate that the sampled peatland was used for hay production.

At Grisavad there was a small-scale excavation close to the sampling site, which focused on the visible remains of a 19th century croft (figure 12). The excavation revealed no medieval remains although the pollen diagram shows that there had been a farm somewhere in the vicinity at that time (A. Knarrström 2003). The arable fields of the late croft were situated very close to the peatland (only 20 metres from the coring point), and they have resulted in a strong influx of cereal pollen in the upper part of the diagram (see the chapter *19th century crofts* below). The cereal signal of the medieval farm is weaker, which probably indicates that it was situated a bit further away from the sampling point than the later croft.

The next site is Värsjö Utmark. It is situated on a till plateau at an altitude of 120–130 m above sea level, which is more than 20 m higher than Grisavad and Östra Ringarp. The pollen diagram from Värsjö Utmark shows a similar early-medieval development as those sites but slightly delayed (figure 13). It starts with temporary cultivation of slash-and-burn type in the 9th and 10th centuries, reflected in some pollen grains of *Secale* (rye) and a peak in microscopic charcoal. The landscape remained forested until the 12th century, when a more open cultural landscape was created. From that time onwards the pollen diagram shows a continuous cereal record and also high percentages of Poaceae undiff. (grass). The major changes in the diagram dated to the 12th century are interpreted as reflecting the establishment of a local farm. The cereal record and the increase in Poaceae undiff. are matched by a decrease in *Fagus* (beech), which indicate that mainly beech forest was cleared to give way to open land. The peak in microscopic charcoal at the

12th century level suggests that fire was used in the initial clearing of the beech forest. The interpretation is supported by the fact that macroscopic pieces of charcoal, which were found beneath an ancient clearance cairn close to the sampling site, were identified as beech and radiocarbon-dated to the 12th century.

An archaeological excavation at Värsjö Utmark revealed late remains of settlement and agriculture, but also some medieval ones (figure 14). In addition to the medieval clearance cairn mentioned above a charcoal kiln and some tar pits were dated to the 13th–14th centuries (A. Knarrström 2004). Furthermore an iron production site situated only 800 m from the Värsjö Utmark site was dated to the same time period (Forenius et al. 2005, Strömberg in prep.).

We may conclude that after some temporary cultivation beech forest was cleared and a farm was established at Värsjö Utmark in the 12th century. The farm grew rye (*Secale*), barley (*Hordeum*), and some wheat (*Triticum*), and it had open grasslands used for grazing and hay mowing. The arable fields were certainly stone cleared, although it is difficult to separate the medieval clearance cairns from the later ones (only two of the clearance cairns were radiocarbon-dated). The agricultural economy was complemented by the production of charcoal and tar, and possibly the farm was also involved in the nearby iron production. It shows that the forest was a resource utilized by the medieval farmers, although it is difficult to quantify its economic importance in relation to agriculture.

The final pollen site to present is Bjärabygget, situated in the northeastern part of the investigation area only six kilometres from the border between the provinces of Scania and Småland, which until 1658 was the national border between Denmark and Sweden. Bjärabygget has a marginal setting with large peatlands and is situated at approximately the same elevation as Värsjö Utmark. The Bjärabygget diagram shows the local establishment of a medieval farm, with a similar process as on the other three sites, but with a rather late establishment, even later than at Värsjö Utmark.

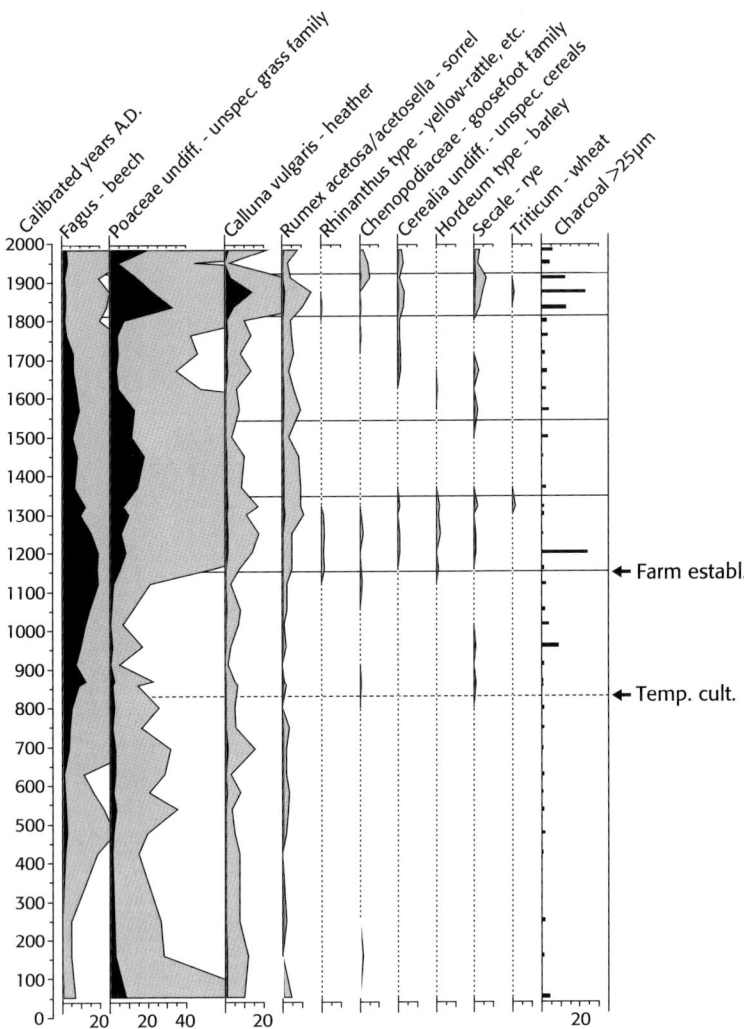

Fig. 13. Pollen diagram from Värsjö Utmark with selected taxa. In addition, the amount of microscopic charcoal in the pollen samples is presented (% of pollen sum). The levels interpreted as reflecting the introduction of temporary cultivation and the earliest farm establishment are indicated. (For details see caption to fig. 9.)

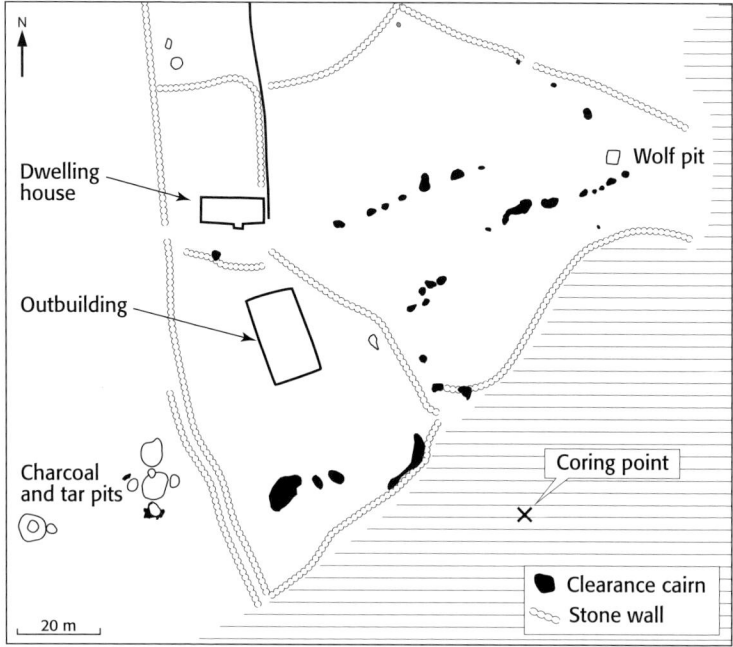

Fig. 14. Map of the Värsjö Utmark site with the remains of an abandoned croft. Hatching indicates peatlands. The coring point for pollen analysis is indicated.

The cereal record starts with a few pollen grains of *Secale* (rye) dated to the 10th and 11th centuries (figure 15). The weak and discontinuous cereal record together with the low frequencies of *Rumex acetosa/acetosella* (sorrel) indicate clearings for temporary cultivation, probably by a farm at some distance. At the same level as the first *Secale* grain the diagram shows an increase in *Fagus* (beech), which may indicate that this small-scale disturbance favoured the establishment of a beech forest (cf. Björkman 1996a).

Starting at a level dated to the 13th century the diagram shows a continuous cereal record, mainly of *Secale*. At the same level there is an increase in Poaceae undiff. (grass), *Calluna* (heather) and in, for instance, *Rumex acetosa/acetosella*. These

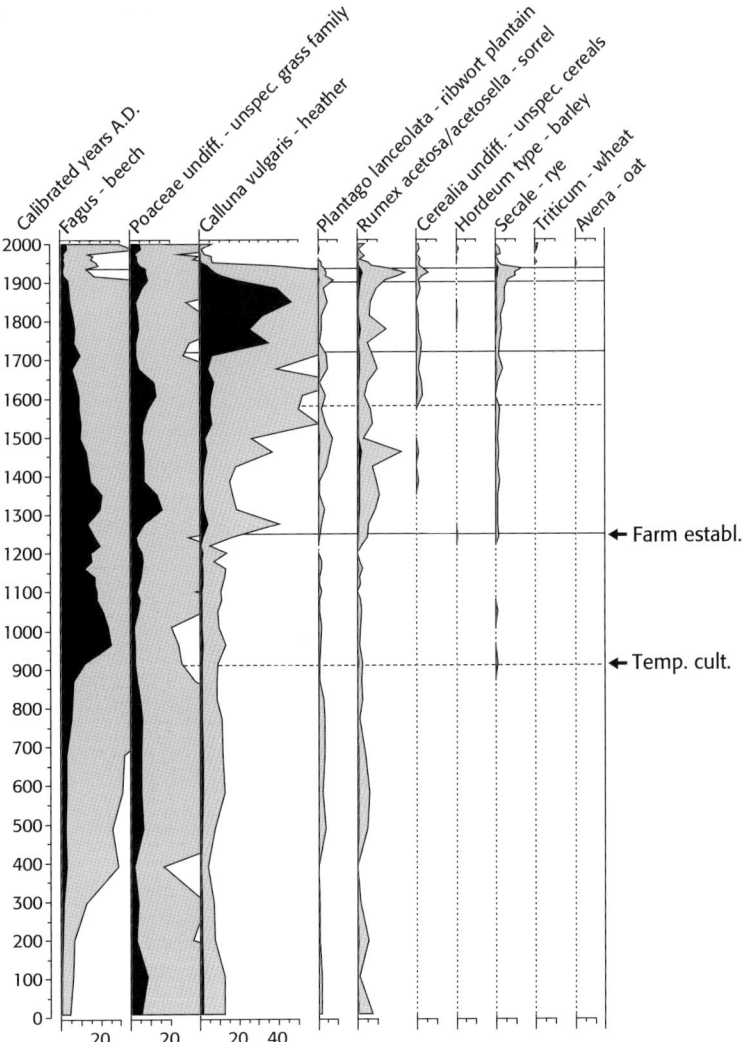

Fig. 15. Pollen diagram from Bjärabygget with selected taxa. The levels interpreted as reflecting the introduction of temporary cultivation and the earliest farm establishment are indicated. (For details see caption to fig. 9.)

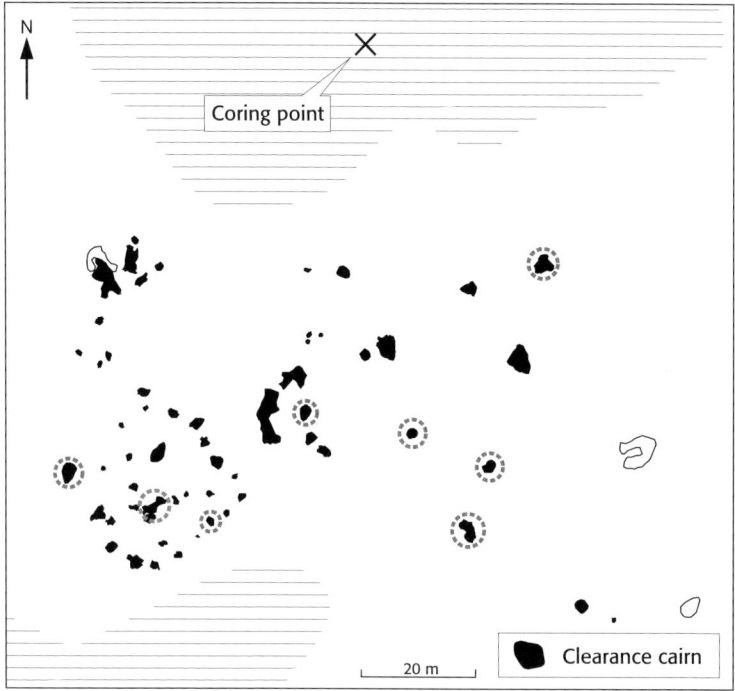

Fig. 16. Map of the Bjärabygget site with ancient clearance carins. Hatching indicates peatlands. The coring point for pollen analysis is indicated. Radiocarbon-dated clearance cairns are marked with a broken circle (see fig. 17).

changes dated to the 13th century are interpreted as reflecting the establishment of a farm nearby. According to the pollen data this medieval farm grew rye and some barley (*Hordeum*), and had grasslands for pasturage and probably also for hay production.

Close to the sampling point at Bjärabygget were the visible remains of ancient agriculture, such as clearance cairns (figure 16), which were subject to archaeological excavation within the project (Linderoth 2004, Lorentzon in prep.). (In connection with the agricultural remains there were also stone foundations of buildings, but they could not be excavated as they

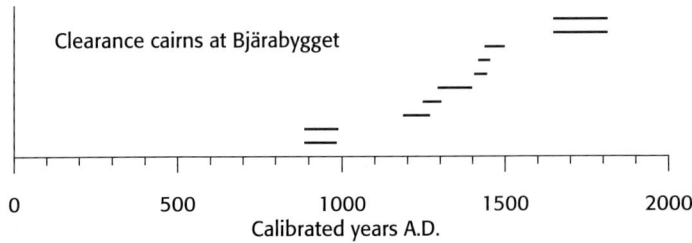

Fig. 17. Radiocarbon dates (calibrated one-sigma intervals) from clearance cairns at Bjärabygget (see fig. 16).

were situated outside the area for exploitation.) The clearance cairns were carefully investigated and altogether ten macroscopic charcoal pieces from eight different clearance cairns were radiocarbon-dated. The results are presented in the diagram in figure 17. As evident from the diagram two charcoal pieces were dated to the 10th century, that is the Viking Period, six were dated to a period which lasted from the 13th to the 15th century, that is the Middle Ages, and finally two (both from the same clearance cairn) were dated to the 17th or 18th century. Of the two early dates from the Viking Period, one is from a clearance cairn that also gave a date to the late Middle Ages. The older date is probably not connected to stone clearance but may nevertheless reflect clearing by fire for agricultural purposes.

The radiocarbon dates show that most of the clearance cairns on the site originate in medieval cultivation, or more precisely from the 13th to the 15th century. The clearance cairns were situated only about 50 m from the coring point and can with certainty be connected to the medieval cultivation reflected in the pollen diagram. Note in particular that the continuous series of dated clearance cairns start at the same time as the continuous record of cereals in the pollen diagram. Hence there is a striking similarity between these two independent sources, showing the establishment of permanent cultivation on the site in the 13th century. Furthermore the two early charcoal pieces from beneath clearance cairns, which were dated to the Viking Period, match the first weak appearance of *Secale* in the pollen diagram, interpreted as temporary cultivation.

Taken together, the pollen and archaeological data show that the medieval expansion at Bjärabygget started with clearances by fire and temporary cultivation during the 10th and 11th centuries. In the 13th century permanent cultivation in stone-cleared fields was introduced, which reflects the establishment of a farm in the local area. At the same time grazing was intensified and pastures expanded.

The results from Bjärabygget give us a picture of the local development on that specific site, but they also give us a valuable key to the interpretation of pollen diagrams in general. The interpretation of cereal pollen is always difficult and in particular when it comes to quantification of the size of arable fields, etc. Attempts have been made to estimate the average pollen production of cereals (e.g. Broström et al. 2004), but it has also been shown that most cereal pollen is not dispersed during flowering but rather in connection with harvest (Vuorela 1973), and therefore we may expect that their pollen dispersal is to a high degree dependent on agricultural technique. This makes the interpretation of cereal records difficult, in particular as, for example, harvesting and threshing techniques have changed through time.

In this situation archaeological data may be a valuable contribution. In all the local pollen diagrams presented in this study, continuous but weak cereal records have been interpreted as reflecting local cultivation (where local means within one or possibly a few hundred metres). Such low cereal percentages could perhaps be interpreted as the result of long-distance pollen transport from arable fields situated far away, in particular as cultivation in connection with later farms on the same sites resulted in much stronger cereal signals. What is interesting about Bjärabygget in this respect is that the weak but continuous cereal record in the pollen diagram, dated to the Middle Ages, is accompanied by a series of clearance cairns of the same age situated only about 50 m from the coring point. It shows that local medieval cultivation may result in a very weak pollen signal, and, the other way around, that a weak cereal record in a pollen diagram indeed may reflect local cultivation. Obviously pollen dispersal from medieval arable fields was very limited, at least in the study area. As it is hard to believe that the medieval arable fields were much smaller than the ones cultivated by later crofts and other small farms in the same areas, the differences in cereal representativeness may perhaps be explained by differences in harvesting techniques. During the Middle Ages cereals were harvested by cutting the ears with a sickle (e.g. Myrdal 1997, Poulsen 1997), while during later times the straw was cut close to the ground using a scythe. It is easy to imagine that the careful sickle harvesting during the Middle Ages resulted in less pollen dispersal than the later technique of using scythes, and it may to some degree explain the relatively weak cereal records of the medieval sections in the pollen diagrams.

After this methodological comment and the site-by-site presentation above, we may now summarize the interpretations from the local pollen diagrams in a regional picture of the medieval expansion. At all the sites the first signs of agricultural expansion (except for the introduction of wood pasturage much earlier) were small-scale clearings and temporary cultivation, which preceded permanent cultivation by some centuries. The phase of temporary cultivation started in the early Viking Period, approximately A.D. 800, in the southwestern part of the area and about a century later in the northeastern part. In most other pollen-analytical studies, single cereal pollen grains at levels where there are few other agricultural indicators are not paid much attention. However, by the consistency of the results from the four diagrams in this study the weak cereal records together make up a convincing pattern which gives us new insights into the expansion process. It shows that the medieval establishment of farms and permanent cultivation were preceded by temporary cultivation and small-scale forest clearing which had started two or three centuries earlier. What is also interesting to note is that – with a possible exception of Östra Ringarp – the cereal records show no continuity from the Viking Period cereal record into the medieval one. On the contrary there seems to have been a break in cultivation that lasted for at least a century.

The weak cereal record of the Viking Period probably reflects outland cultivation carried out far away from settlements. It was a small-scale, temporary cultivation in forest clearings and was probably intended only to be so. However, an alternative interpretation, which cannot be completely ruled out, is that this temporary cultivation reflects the introductory phase of an agriculture and settlement expansion that for some reason was never completed or fulfilled. This is a thrilling thought, but unfortunately it cannot be proved or disproved by the data at hand.

Whatever the purpose of the early clearings, we may conclude that the conditions for agriculture at sites chosen for farm establishment during the Middle Ages were not unknown to the farming community – their suitability had already been tested by temporary cultivation during the Viking Period. We do not know if the people who made these early clearings came from the same settlements as the later ones who established permanent farms, or if they were in any way related. If there was no direct relationship or continuity in this respect, the reason why sites that had once been used for temporary cultivation were later regarded as suitable for settlement and permanent cultivation may have been that they showed some quality due to the early clearings, such as, for instance a more open forest structure, which attracted the settlers.

As mentioned above the temporary cultivation started slightly later in the northeastern part of the investigation area than in the southwest. If we look at the next phase – the establishment of local farms practising permanent cultivation – this time/space gradient is even more pronounced. Moving from the southwest to the northeast, which also represents a gradual increase in altitude from 100 m to about 130 m above sea level, the farm establishment is dated to the 11th century at Östra Ringarp and Grisavad, to the 12th century at Värsjö Utmark, and to the 13th century at Bjärabygget (figure 18). Hence

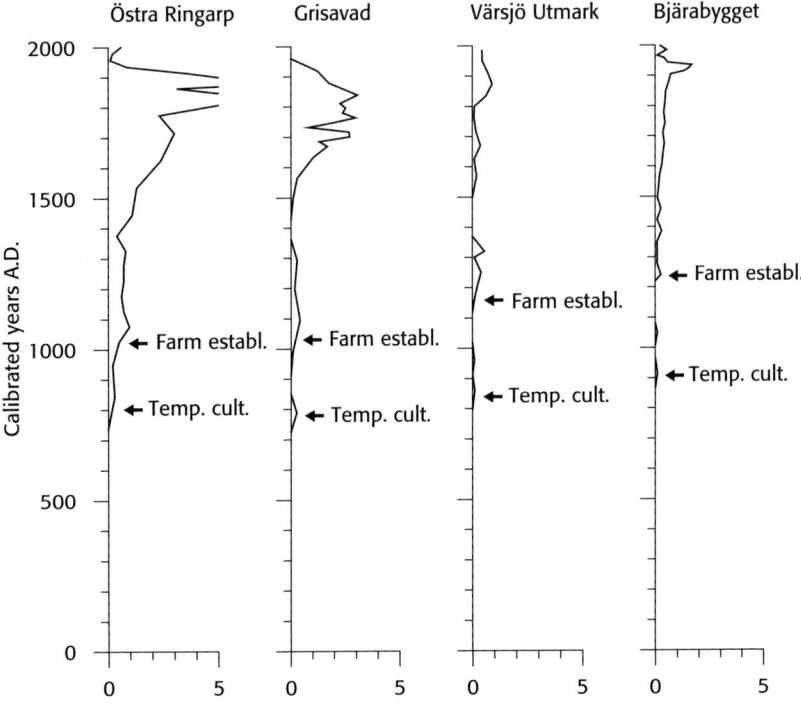

Fig. 18. Summary graphs for all cereal pollen [i.e. Cerealia undiff. (un-spec. cereals), Hordeum type (barley), Secale (rye), Triticum (wheat), and Avena (oat)] from each site. The graphs show percentages of total pollen. Levels interpreted as reflecting the introduction of temporary cultivation and the earliest farm establishment are indicated.

the colonization process in the study area may be envisaged as a moving frontier from southwest to northeast, gradually reaching higher elevations. An interesting result is that the medieval expansion was obviously still going on during the 13th century. What is also interesting to note is that the farm at Bjärabygget was situated relatively close to the national board-er between medieval Denmark and Sweden. Even though the representativeness for a larger area is unclear, the relatively late establishment at Bjärabygget may illustrate how the uninhabit-ed borderland was gradually colonized – a process which also

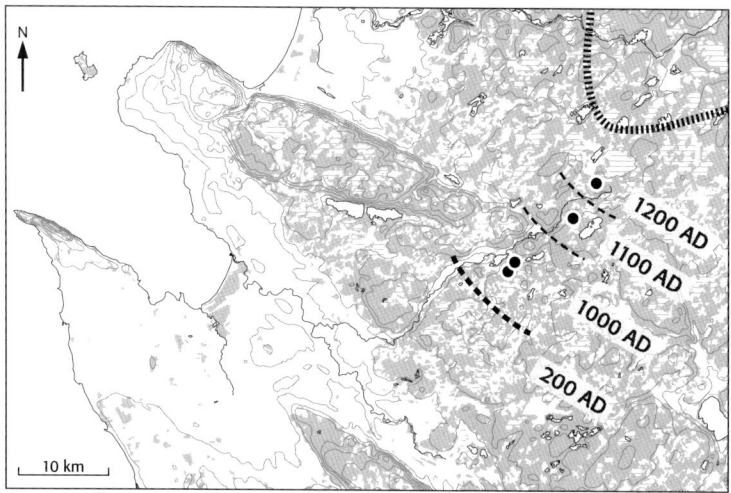

Fig. 19. Map showing the different steps of the agricultural colonization of northern Scania. The map is based on dated clearance cairns (Lager-ås et al. 2000) and pollen diagrams (present study).

may have contributed to the need for an unambiguous and more distinct border.

This picture of a settlement frontier moving from southwest to northeast gradually reaching higher elevations becomes even more striking if we broaden the perspective in both time and space, and also include other source material than pollen data. Thanks to a series of archaeological and palaeoecological investigations in connection with the construction of the new E4 motorway, not only within the present project but also in connection with a previous road section, we can follow the upland expansion of permanent settlement and cultivation all the way from the fertile plains in the southwest (figure 19). Down on the plains the richness of prehistoric remains from all periods shows a settlement continuity from the Neolithic onwards. The first step up on the upland was taken in the Roman Iron Age, about A.D. 200, according to the results of the investigation and radiocarbon dating of large areas with clearance cairns (Lagerås et al. 2000). Among the clearance cairns there were also the remains of long houses, iron production, etc., from the same time. This Roman expansion reached only about five kilometres up on the upland along the investigated gradient. The next step of expansion was taken considerably later, during the Middle Ages, when agriculture and settlement establishment reached Östra Ringarp and Grisavad in the 11th century, Värsjö Utmark in the 12th century, and Bjärabygget in the 13th century (figure 19).

Obviously the colonization of the uninhabited areas of northern Scania was a step-wise process, which gradually reached further up into the upland. New farms were not established in completely unknown areas, but rather within the activity area of earlier settlement, that is within a zone used for wood pasturage, temporary cultivation, and other outland activities. A plausible interpretation is that the new settlers were still tied in one way or another to their parent settlements. They were socially and perhaps also economically linked to the settlements they came from and they were still part of the same society. The forested uplands offered new possibilities for a diverse economy, based not only on agriculture, but also on, for example, iron production, coaling, forestry, hunting, etc. However, the settlers came from an agricultural society, and we may therefore regard agriculture as the fundamental basis for the medieval expansion even in an upland area such as northern Scania.

The medieval land-use and cultural landscape

After the interpretation of the colonization process we may now try to combine different sources of information and characterize the medieval agricultural land-use in the investigation area. To start with, however, it can be concluded that the Middle Ages from many points of view is still an enigmatic period, not least when it comes to land-use. The fact that the Middle Ages by definition belong to the historic period, and not to the prehistoric one, may fool us into believing that medieval landscapes, land-use techniques, etc., should be familiar to us and hence should be easy to interpret. Sometimes also a too simple division is made, for example, between prehistoric and historic settlement patterns, or between prehistoric and historic land-use systems, which gives the impression that historic conditions known from later periods may easily be extrapolated back to the Middle Ages. In Swedish archaeology such a static view of the landscape and settlement development of the historic period is quite common. Clearance cairns and other agricultural remains that according to maps from the 18th or 19th centuries are situated in non-cultivated outlands are interpreted as being prehistoric simply because they do not fit into the late historic settlement and land-use pattern. Several investigations have shown, however, that many such agricultural remains in outland areas may be medieval, and that there have obviously been many changes within the historic period (e.g. Vestbö-Franzén 1997, Lagerås 2002a, Ericsson 2004).

Also in the investigation area, in the uplands of northern Scania, the landscape has undergone several changes since the Middle Ages. Hence, the information that we have got from written sources from the 16th century onwards, for example on agricultural production, cannot be extrapolated to the Middle Ages without caution (Skansjö 1997a). Interpretations of the local land-use and cultural landscape have to be based primarily on palaeoecological data together with the radiocarbon dates of some agricultural remains. The local interpretations may also be compared to the general picture of Swedish medieval agriculture as presented by Myrdal (1985, 1999).

To begin with, the cultivation of cereals in the investigation area during the Middle Ages is reflected first of all in the occurrence of cereal pollen. *Hordeum*-type pollen was found in the medieval sections in all the diagrams (figures 9, 11, 13, 15). This pollen type includes different species of barley as well as one wheat species (einkorn *Triticum monococcum*), which are not possible to distinguish based on pollen analysis alone. However, of the different species and varieties included in this pollen type, macrofossil analyses of charred cereal grains have shown that the only one that was regularly cultivated in Sweden during the Middle Ages was hulled barley (*Hordeum vulgare* var. *vulgare*) (Engelmark & Viklund 1990, Engelmark 1992), and all others were absent or very rare. We may therefore conclude that the identified *Hordeum*-type pollen grains reflect local cultivation of hulled barley.

Also *Secale* pollen was found in the medieval sections in all the diagrams of the area. The only possible *Secale* species is rye (*Secale cereale*), which obviously was cultivated at all the investigated sites.

In addition to *Hordeum* type and *Secale*, some pollen grains of *Triticum* have also been identified at three of the four sites. It is missing in the medieval layers at Bjärabygget, which is the most northeastern site. *Triticum* covers several wheat species that are difficult to separate, but according to macrofossil analyses the only wheat species that was cultivated regularly during the Middle Ages in southern Sweden was bread wheat (*Triticum aestivum*) (Engelmark 1992). The *Triticum*-pollen grains probably indicate cultivation of this species.

In the pollen diagrams there is also a taxon called Cerealia undiff., which mainly includes cereal pollen that could not be identified to genera, due to poor preservation, folding, etc. However, it may also include pollen grains of a few wild grass species, which, similarly to cereals, have large pollen grains, in particular floating sweet-grass (*Glyceria fluitans*), lyme-grass (*Leymus arenarius*), and couch (*Elytrigia repens*). If lyme-grass or couch grew in the area it was certainly as weeds in arable fields, which makes them indicators of cultivation almost in the same way as cereals. Floating sweet-grass may be more problematic as it grows on natural wet soil and need not to be connected to cultivation or any cultural land. This possible source of error seems, however, to be of no importance in the present study, as the graph for Cerealia undiff. in all the diagrams follows the graphs for the cereals identified with certainty. Most probably many of the pollen grains referred to as Cerealia undiff. are in fact barley.

Another cereal that is possible to distinguish in pollen analysis is *Avena*, which includes the cultivated species oat (*Avena sativa*) and the weed wild-oat (*Avena fluitans*). In the present study no pollen of *Avena* was found in layers dated to the Middle Ages, but it was found in layers from the last three hundred years. Even though the possibility of small-scale cultivation also during the Middle Ages cannot be excluded based on negative evidence, the results indicate that oat was not an important crop until later. Studies based on written sources have shown that oat was an important crop particularly in western Sweden at least from the 16th century onwards (Söderberg & Myrdal 2002), and according to macrofossil analyses this seems to have been the case as early as the Viking Period (Viklund 2003). The investigation area in northern Scania is on the rim of western Sweden, but the results indicate that it was not embraced by the western tradition of oat cultivation, at least not during the Middle Ages.

To sum up, hulled barley, rye, and bread wheat were the three major crops in the area during the Middle Ages. The quantitative relationship between these different crops is difficult to judge from the pollen data, partly because of the low frequencies and partly also because of differences in pollen production, dispersal, etc. Most cereals are autogamous and have poor pollen dispersal as much of their pollen remains in the hulls (Behre 1981). The exception is rye, which is an allogamous species with higher pollen production and more efficient wind dispersal. In relation to the other cereals, rye is probably over-represented in pollen diagrams. Having this in mind, the results indicate that barley and even wheat may have been of similar importance to rye during the Middle Ages in the study area, except for at Bjärabygget in the northeast where there was a very strong dominance of rye.

The picture of medieval crops presented here, which is based on pollen analysis, fits in relatively well with the one given by Myrdal (1999), which based his interpretation mainly on written sources from the 14th and 15th centuries in Middle Sweden. According to him, barley was the dominating crop followed by rye, while wheat and oat played a minor role. A similar picture has emerged from macrofossil analyses of charred cereals from southern Sweden (e.g. Hjelmqvist 1992). There were regional and local variations, however, as exemplified by the importance of oat in western Sweden as mentioned above.

In addition to cereals the pollen diagrams from the present study provide evidence for two other cultivated species during the Middle Ages, namely hemp (*Cannabis sativa*) and flax (*Linum usitatissimum*). Hemp was cultivated at one of the sites, Östra Ringarp, from the initial farm establishment in the 11th century throughout the Middle Ages and all the way up until approximately the early 19th century. Pollen of hemp may be difficult to distinguish from those of hop (*Humulus lupulus*), and they are both included in the pollen taxon *Cannabis* type. However, the cultivation of hop only involved female plants (Hansson 1996), which of course do not produce pollen. Some male plants may have been around, but it would still not explain the high percentages from Östra Ringarp. The high values could only be explained by the cultivation of hemp, which is a plant that produces a great amount of pollen (Subba Reddi & Reddi 1986). Furthermore, the cultivation of hemp in contrast to hop seems to have focused in particular on male plants as they produce the best fibres (Edwards & Whittington 1992, Mercuri et al. 2002).

Hemp is not a native plant to Sweden but it was introduced during the Iron Age and it was cultivated in many regions during the Middle Ages. It has leaves and flowers with narcotic properties and also oil-rich seeds, but in early times it was cultivated first of all for its fibres, which could be used to produce ropes, cloth, fishing nets, etc. After harvest the fibres were extracted through a process called retting, in which bundles of stems were deposited in standing water for six or more weeks, after which they were dried and twisted to make ropes, etc. (Schofield & Waller 2005). High pollen percentages of *Cannabis* type are often interpreted as evidence of hemp retting in the actual lake sampled for pollen analysis, both in Swedish studies (e.g. Regnéll 1989, Gaillard & Göransson 1991) and abroad (e.g. Bradshaw et al. 1981, Mercuri et al. 2002, Rasmussen & Anderson 2005, Schofield & Waller 2005). The high percentages

at Östra Ringarp may in the same way be interpreted as originating from retting. However, even though the values are high
– about 5 % of the pollen sum – they are much lower than in
several other studies, in which they sometimes reached more
than 30 %. Therefore, and having the rich pollen production
of hemp in mind, the *Cannabis* record at Östra Ringarp may
not necessarily originate from retting, but simply from hemp
fields close to the site. In either case, however, it reflects local
cultivation.

In the medieval section in the diagram from Östra Ringarp
there is also one pollen grain of *Linum usitatissimum* (flax).
One pollen is not much, but this entomogamous (insect-pollinated) species is a very poor pollen producer, which never
shows a strong signal in pollen records. Usually only one or a
few pollen grains are found, even in samples within the cultivated areas (Behre 1981). The single flax pollen from Östra
Ringarp is therefore interpreted as an indication of local cultivation, even though it is not possible to conclude anything
about its importance.

It is worth noting that the medieval farm at Östra Ringarp
cultivated both hemp and flax, as both these crops were grown
for the fibres, and were processed partly in the same way. A
similar relationship – that pollen of flax and *Cannabis* type occur at the same levels – has been shown also by some other pollen-analytical studies (e.g. Peglar et al. 1989, Regnéll 1989,
Gaillard & Göransson 1991). In one Danish study, large numbers of macrofossil seeds and capsules of flax in medieval lake
sediments were interpreted as the remains of retting. Pollen
analysis of the same layers revealed abundant *Cannabis*-type
pollen but only three pollen grains of flax (Rasmussen 2005,
Rasmussen & Anderson 2005). It illustrates the weak pollen
signal of flax and its poor representation in pollen spectra, even
when it has obviously been cultivated in the vicinity and retted
in the actual lake. Therefore it is justified to interpret a single
flax pollen grain, in particular when it is found in the same layers as *Cannabis* type, as an indication of local cultivation. We
may conclude that the farm at Östra Ringarp during the Middle

Ages and later was involved in fibre production for ropes and textiles – a specialization that we do not find indications of at any of the other three investigated sites. (However, flax pollen was found at Grisavad in much younger layers, dated to the 19th century; see the chapter *19th Century Crofts.*)

After this résumé of positive local evidence of medieval crops it should be mentioned that it is not possible to prove all crops through pollen analysis (table 1). From early written sources and also from macrofossil analyses (Myrdal 1985, Engelmark 1992) we know for example that field pea (*Pisum sativum*), horse bean (*Vicia faba*), and turnip (*Brassica rapa*) were cultivated in both Denmark and Sweden during the Middle Ages. Field pea and horse bean belong to the Fabaceae family while turnip belongs to the Brassicaceae family. They are all poor pollen producers and their pollen types are not possible to distinguish from other taxa within the same families. Hop (*Humulus lupulus*) was also cultivated during the Middle Ages but due to the preference for female plants, and also due to the similarity between the pollen of hop and hemp (see above), its cultivation may be difficult to prove or disprove by pollen analysis. It is possible that some or all of these crops were also grown in the investigation area, in spite of the lack of pollen-analytical evidence. Therefore, to conclude, the list of crops cultivated in northern Scania during the Middle Ages according to the pollen analyses presented in this study – that is rye, barley, wheat, hemp, and flax – make an important contribution to our knowledge but it probably does not give the full picture of the diversity of medieval crops.

If we now leave the crops and look at the arable land, the medieval cultivation identified in the pollen diagrams took place in stone-cleared fields. This is evident from radiocarbon-dated clearance cairns throughout the investigation area – altogether twelve clearance cairns from six different archaeological sites were dated to the Middle Ages, that is to the time period A.D. 1050–1500 (figure 7). The dated material was charcoal, mainly from beech (*Fagus*), sampled from the bottom layers of the cairns. It probably originated from burning of

Table 1: Some cultivated plants and their representation in pollen diagrams. Representation in this context means not only that a pollen taxon may be present in the samples but also that it is possible to distinguish from other taxa.

Strong representation	Weak representation	No representation
Rye (*Secale cereale*)	Barley (*Hordeum vulgare*)	Field pea (*Pisum sativum*)
Hemp (*Cannabis sativa*)	Wheat (*Triticum* sp.)	Horse bean (*Vicia faba*)
	Oat (*Avena sativa*)	Turnip (*Brassica rapa*)
	Hop (*Humulus lupulus*)	Potato (*Solanum tuberosum*)
	Flax (*Linum usitatissimum*)	

twigs etc. when a plot was cleared and prepared for cultivation.

Many of the areas with medieval clearance cairns give an amorphous and irregular impression, while others show a more advanced structure with strip fields. This structure is recognizable in ancient maps and sometimes also in the field, by fragments of stone walls, terraces, or earth banks. Among the four pollen-analysed sites in the study, strip fields were identified only at Östra Ringarp, but they have been identified at several other sites within the investigation area (Connelid 2003), and they are also known from other parts of southwestern Sweden (e.g. Mascher 1992, Widgren 1997, Connelid 2002, 2004).

The strip fields probably reflect a division of the agricultural land among different farmers, but such an interpretation seems at first to be in conflict with the fact that the medieval settlement pattern in the investigation area was dominated by single farms. During the 17th century, from which we have relatively good historical data, the vast majority were single farms while only 20 % made up small hamlets consisting of three or four farms (Skansjö 1997a). In contrast to this, the settlement structure on the agricultural plains in southern Scania was dominated by large hamlets or villages, with about 15 farms on average (Dahl 1942). However, a complicating factor when interpreting the settlement structure from early registers is that a "farm" not only refers to a settlement unit but also to a tax-paying unit, which may not necessarily be the same. In the investigation area, the cadastral registers from the 16th century, as well as the Swedish Civil Survey conducted after the conquest of Scania in the 17th century (Sw: *Jordrevningsprotokoll*), mention by name several different farmers at each single farm. This is particularly true for farms with a presumed early-medieval origin (Skansjö in prep.). Hence, it seems as if some single farms functioned as hamlets with several different farmers, even though they are called farms and not hamlets in the documents. This probably explains why they developed a strip field division of the land. Östra Ringarp is a typical example. It not only shows a strip field pattern – according to a cadastral register (Sw: *Decimantjordeboken*) from 1651 it was a farm (in the meaning of a tax paying unit) with eight different farmers. Possibly there were several different farmers at Östra Ringarp (then only called Ringarp) during the Middle Ages as well.

As mentioned above, Östra Ringarp is the only one of the four pollen-analysed sites where we have found evidence of strip fields. Together with evidence of medieval hemp and flax cultivation inferred from pollen data, and also local iron production, it gives the impression of having been a well-established medieval farm with a diverse economy and probably several farmers. Other sites such as, for example, Bjärabygget, with its rye cultivation and simple clearance cairns, give a more modest impression. In conclusion the medieval farms and the land-use systems in the area were not all uniform, but instead showed diversity, ranging from small single farms to large hamlet-like ones with a more complex economy, organization and land-use.

In addition to the rather intense cultivation in strip fields and among clearance cairns discussed so far, there may also have been extensive crop growing, for example slash-and-burn cultivation. The earliest written document on slash-and-burn cultivation in southern Sweden is from the 14th century (Vestbö-Franzén 2004), and in the 15th and 16th centuries it was widespread in most forest regions (Myrdal 1999a). In the present study, single cereal pollen grains from some centuries before the early-medieval farm establishment were interpreted as temporary cultivation, possibly of the slash-and-burn type, and it cannot be excluded that such extensive cultivation was also practised in outlands after the colonization. This is however difficult to prove or disprove. The most common crops in this type of land-use were rye (*Secale cereale*) and turnip (*Brassica rapa*), of which only rye can easily be traced by pollen analysis. As rye was also grown in infields, single pollen grains from temporary clearings may be hidden in the *Secale* record, which certainly is dominated by pollen from continuous cultivation in stone-cleared fields.

So far, focus in this section has been on cultivation and crop growing, but animal husbandry was equally or probably more important. The pollen diagrams reflect open or semi-open grasslands during the Middle Ages at all the investigated sites. Most of these grasslands were certainly pastures, but some may also have been used for winter fodder production through mowing. In the pollen diagrams these pastures and meadows are reflected first of all in high percentages of Poaceae undiff. (grass), and also in a great diversity of herbs, for instance *Plantago lanceolata* (ribwort plantain), *Rumex acetosa/acetosella* (sorrel), *Potentilla* type (tormentil, cincuefoil, etc.), and *Filipendula* (dropwort, meadowsweet). Grasses of different species certainly dominated, but it should be noted that most other herbs are entomogamous (insect-pollinated) and are therefore underrepresented in pollen diagrams. The low frequencies and sporadic occurrence of some of this herb pollen are difficult to interpret and they may be regarded as qualitative indicators rather than quantitative ones. Examples of pollen taxa that are represented only by a few pollen grains, but nevertheless contribute to the picture of the medieval grassland vegetation, are *Rhinanthus* type (yellow-rattle, eyebrights), *Trollius europaeus* (globeflower), and *Circium* (thistles).

The possibility of distinguishing mowed hay meadows from grazed pastures in pollen spectra has been discussed by several authors (e.g. Greig 1983, Gaillard et al. 1994, Lagerås 1996a). Even though stress caused by mowing is similar to stress caused by grazing, these two types of land-use may result in different plant composition (Ekstam et al. 1988). In pastures there is often a larger proportion of thorny, stinky and bitter plants than in hay meadows (e.g. thistles), simply because this type of defence works on grazing animals but not on scythes. Pastures may also have a larger proportion of trample-resistant plants. In the long run, meadows may become more nutrient-depleted, due to hay collecting, while in pastures more nutrients are returned to the system as dung. Finally, the fact that the grazing season begins some months prior to hay mowing, which in southern Sweden is traditionally in the middle of July or later, gives some plants better possibilities to flower in hay meadows than in pastures. From these examples it is evident that grazing and mowing result in somewhat different plant compositions, but in practice much of this difference is smoothened out, partly because most meadows were grazed in late summer and fall after mowing.

A problem when it comes to pollen analysis is that the indicator species that may distinguish meadows from pastures are often poor pollen producers, or they have pollen that is hidden in pollen types that include many taxa, for example Poaceae undiff. (grass), Rosaceae undiff. (rose family), Asteraceae Lactucoideae (dandelions, hawk's-beards, etc.), *Aster* type (groundsel, colt's-foot, etc.), and Ranunculaceae undiff. (buttercup family). In the pollen diagrams from the investigation area there is no certain evidence of hay meadows, mainly due to these difficulties, although the pollen grains of, for example, *Rhinanthus* type (yellow-rattle, eyebrights) and *Centaurea nigra* type (common knapweed) may be regarded as positive indicators.

A plant that was absent or at least rare in most pastures at the time is heather (*Calluna vulgaris*). This is an interesting fact as heather totally dominated the pastures in the same area some hundred years later, in particular during the 19th century. (The late expansion of heathlands is discussed in more detail in the chapter *Man-made Heathland: Birth and Decline*.) For some reason, most pastures during the Middle Ages never developed into nutrient poor heathlands in spite of several hundred years of grazing. Obviously grazing was not enough to produce heathland in this region, at least not during the prevailing climatic conditions. The question why heather expanded later is difficult to answer, but it may have been a combination of several factors, such as climatic deterioration (the Little Ice Age), soil degradation, and the increased use of fire. Note that medieval agriculture benefited from the good climate of the Medieval Warm Period (see page 30) and also of the brown earth soils that had been built up for thousands of years in deciduous woodland.

There was however a small-scale expansion of heather as early as the Middle Ages at two of the investigated sites: Grisavad and Bjärabygget. In the diagrams from these sites there is a small but significant increase in *Calluna* pollen dated to the 13th century (figure 20), a rise which may reflect the first establishment of heathland in some restricted areas. It is worth noting that the oldest documented place-name in the area is Örkelljunga (then spelled Øthknælyung), which is mentioned in a letter from 1307. Örkelljunga was the main settlement of the parish, with a church and a royal castle. The first part of the name, *Örkel-*, is related to *öken*, which is the Swedish word for desert, in this context probably meaning uninhabited or deserted. The ending *-ljunga* is related to *ljung*, which is the Swedish name for heather and in this context probably means heathland (Pamp 1988). Hence, the place-name indicates that there was at least some heathland in the vicinity of Örkelljunga at the beginning of the 14th century and probably earlier, and we may for good reason connect this with the rise in *Calluna* pollen dated to the 13th century at Grisavad and Bjärabygget. The distance between Grisavad and Örkelljunga was only three kilometres.

According to the pollen diagrams, the medieval pastures had relatively few shrubs. In particular the absence of *Juniperus* (juniper) pollen is significant, as this taxon normally produces a great deal of pollen that is easy to identify. A few *Juniperus* pollen grains are found in the medieval sections but there is no major expansion until the 16th century and later (figure 21). Some willow (*Salix*) may have grown in pastures on damp

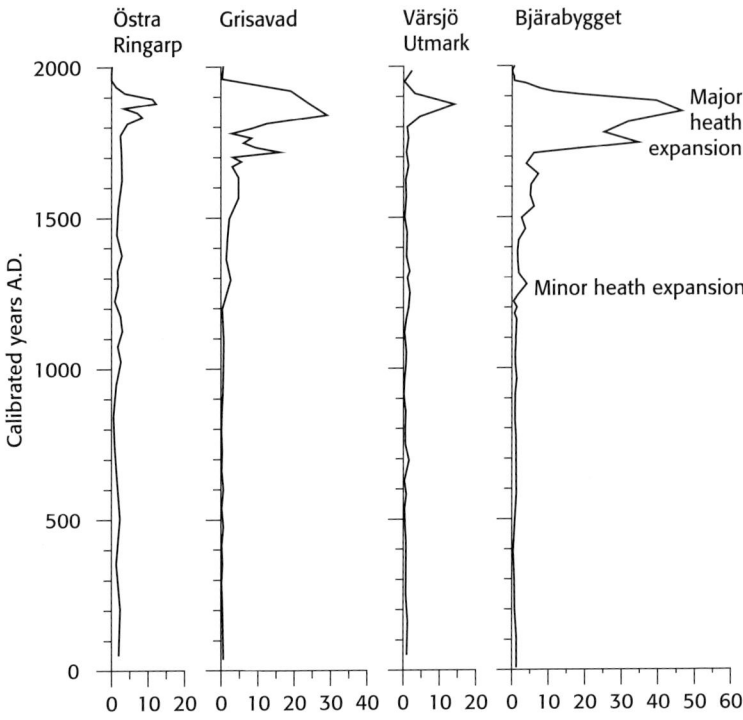

Fig. 20. *Calluna vulgaris (heather) pollen graphs for each site. The graphs show percentages of total pollen.*

soil and bog-myrtle (*Myrica*) in wetlands, but the over-all impression is that shrubs were few. It must be noted, however, that shrubs of the rose family, that is of the genera *Prunus* (blackthorn, etc.) and *Crataegus* (hawthorns) are poor pollen producers and they may have been present even though their pollen has not been detected.

After these rather detailed discussions on different crops, land-use, etc. we may now try to summarize this section by trying to envisage what the medieval landscape in northern Scania looked like:

The forests had been used only for herding and extensive wood pasturage for thousands of years, and during the Viking

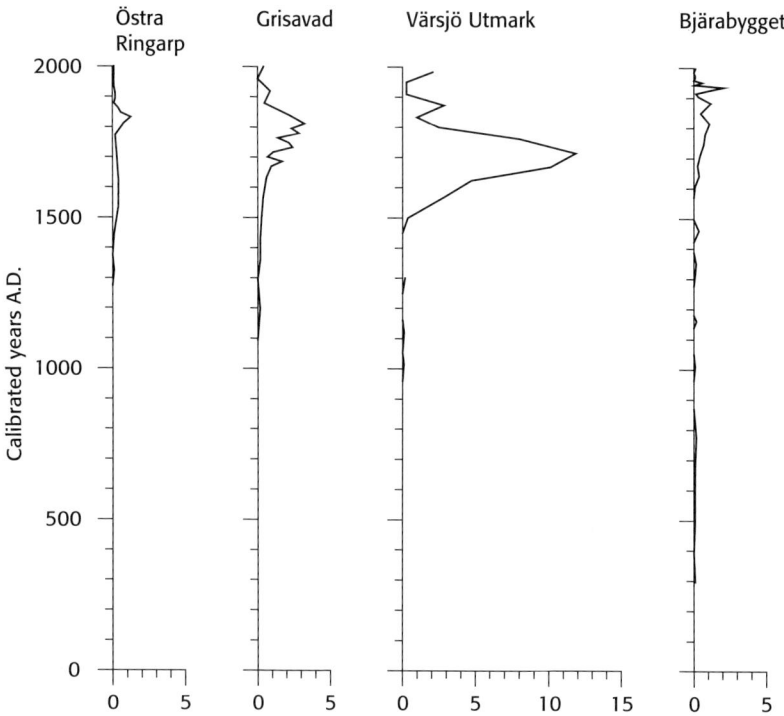

Fig. 21. Juniperus (juniper) pollen graphs for each site. The graphs show percentages of total pollen.

Period – as a prelude to the medieval expansion – the area was also subject to extensive cultivation in temporary clearings. However, it was the colonization and farm establishment in the early Middle Ages that triggered a more profound landscape change and the birth of a diverse cultural landscape in the uplands of northern Scania. This medieval cultural landscape was characterized by a dispersed settlement structure with single farms and possibly a few small hamlets. The farms had stone-cleared fields used for growing cereals and other crops. Rye and barley were the most common cereals and were probably cultivated at every farm. At some they also cultivated wheat, and at a few specialized farms they also cultivated fibre

67

plants such as hemp and flax. The arable of most farms was irregular, while some – probably larger ones with several farmers – had their arable organized in strip fields.

Close to arable and settlement were also open pastures as well as meadows for hay production. Both pastures and meadows had a rich flora characterized by grasses and herbs, and they thus differed from the poor heathlands of later periods. They also had stands of light-demanding trees such as birch and hazel and possibly shrubs of, for instance, hawthorn and blackthorn (but almost no junipers). Trees close to settlement were probably pollarded for leaf fodder although it is not possible to prove this by pollen analysis.

Leaving the arable and the open grasslands that surrounded settlements and moving further out into the outland the landscape was still rather forested. It was a deciduous forest dominated by beech and oak, and in much smaller numbers trees such as lime and hornbeam. In lighter and more open parts of the forest thrived birch and perhaps some aspen and rowan, while alder bordered lakes and fens. There was one coniferous tree species present in the region, namely pine, but it was by large confined to bogs.

The forests were not wilderness, but rather part of the cultural landscape and became so more and more. They provided timber for buildings and firewood, and they were the basis for the production of charcoal and tar (one important consumer of charcoal was the local iron production). The forests were also used for wood pasturage and probably for leaf fodder collection, which, together with the other human activities, certainly affected the forest structure. Grazing and browsing ought to have been most intense in the vicinity of farms, which resulted in a slow gradient from open land around settlement to dense forests far away. Apart from these general conclusions about the forest structure it is difficult to quantify the denseness or openness of the medieval forest in absolute numbers.

LATE-MEDIEVAL DECLINE

·········

Local evidence of agricultural decline and abandonment

So far focus has been on the expansion process and on the agriculture practised by the early-medieval farmers. The four different local pollen diagrams presented here give a rather unanimous testimony of the expansion. They show a time gradient, with an earlier expansion in the southwestern part of the investigation area than in the northeast, but all the sites witnessed farm establishment and the introduction of permanent cultivation within the time period A.D. 1000–1300. They thus tell more or less the same story of early-medieval expansion. However, during the course of the Middle Ages the picture changed and the different farms faced different destinies. We will now look at the late-medieval development site by site. It will turn out that two sites in particular are interesting in that they show indications of farm abandonment in the 14th century. These sites are Grisavad and Värsjö Utmark.

In the diagram from Grisavad a weak but continuous cereal record starts at a level dated to the 11th century, which was interpreted as the time of farm establishment. The farm grew several different crops and had grasslands used for pasturage and probably hay production. Interestingly enough, however, the cereal record shows an abrupt end at a level dated to the 14th century (figure 22). It reflects how cereal growing in the local area ceased, which was the end of a 300-year period of permanent cultivation at the site.

Following the end of the cultivation there are some signs of overgrowing. An immediate response seems to have been an expansion of willow shrubs (*Salix*), while a slower response is shown by the expansion of *Pinus* (pine) and in particular of *Fagus* (beech). Later, when cultivation was introduced again in

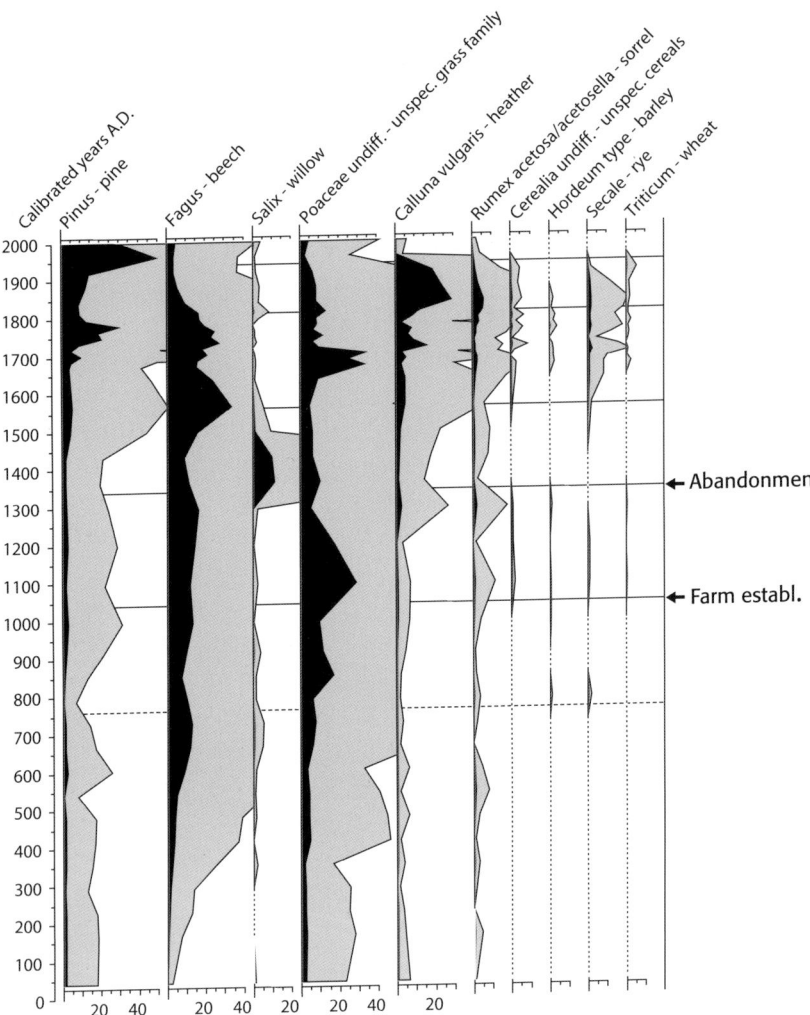

Fig. 22. Pollen diagram from Grisavad with selected taxa. The levels interpreted as reflecting early-medieval farm establishment and late-medieval abandonment are indicated. (For details see caption to fig. 9.)

the late 16th century, reflected in the reappearance of cereal pollen, the diagram shows a sharp decrease in *Fagus* and to some degree also in *Pinus*. This relationship probably reflects how the same fields that were abandoned in the 14th century once again were cleared and cultivated.

In spite of the ceased cultivation and overgrowing of some land, the relatively high percentages of grass pollen (Poaceae undiff.) together with the large number of herb pollen taxa show that there were still some grasslands in the area. Obviously cereal growing ceased while some grazing continued, and the most plausible interpretation is that the local farm was abandoned and that parts of the area were still used for grazing by one or several distant farms. The local land-use thus turned from intensive to more extensive. It is not possible from the pollen data from this site to say if mowing continued or if the hay meadows were also turned into pastures.

As the late-medieval decline has been explained by some authors as the exhaustion of poor soils in marginal areas (see discussion in next section), it should be mentioned that there is a decrease in grass pollen and an increase in *Calluna* (heather) in the Grisavad diagram at a level dated to the 13th century. This change, which happened about a century before the abandonment, indicates that some grazed pastures turned into poor heathland, and thus could reflect soil exhaustion. However, the increase in *Calluna* is rather weak and there was obviously still a great deal of grassland that did not develop into heathland. In particular, if one compares this with the much stronger development of heathland in the 19th century, which is reflected in high *Calluna* percentages in the upper part of the diagram, it seems as if heath formation could not have been a major problem in the Middle Ages.

The other site within this study that shows signs of late-medieval farm abandonment is Värsjö Utmark. In the pollen diagram from Värsjö Utmark the start of a continuous cereal record and a major increase in open-land taxa were dated to the 12th century and interpreted as reflecting the establishment of a local farm. In the same way as in the diagram from Grisavad, the medieval cereal record from Värsjö Utmark ends abruptly at a level dated to the 14th century (figure 23). Cereal growing ceased at this site after about two hundred years of permanent cultivation. Also notable is that some other taxa, which reflect weeds on arable or ruderal land, are present at levels from before the 14th century but are absent after, such as *Urtica* (nettle), *Rumex obtusifolius* type (dock), and Chenopodiaceae (goosefoot family).

In contrast to Grisavad, however, there seems to have been no overgrowing by shrubs or trees on abandoned land at Värsjö Utmark. On the contrary the diagram shows an expansion of Poaceae undiff. (grass) and several other grassland taxa, such as *Plantago lanceolata* (ribwort plantain), *Melampyrum* (cowwheat), *Potentilla* type (tormentil, cinquefoil), and *Ranunculus* type (buttercup). Obviously the area was still grazed. As at Grisavad the most plausible interpretation is that the local farm was abandoned and that the area was used for grazing by other farms situated farther away. There are at least two possible explanations as to why grassland indicators increased when cultivation ceased. One is that abandoned arable fields were transferred to pastures, which meant a total expansion of grassland. The other is that a slightly decreased grazing pressure resulted in richer flowering of many grassland taxa, which in the pollen diagram gives the impression of grassland expansion. However, both explanations may be valid as one does not exclude the other, and they both imply a shift from intensive to more extensive land-use.

As always it is difficult to distinguish hay meadows from pastures, but there are some changes in the herb pollen composition, which may reflect a change in grassland use. In particular *Rhinanthus* type (yellow-rattle etc.) shows an interesting distribution in the pollen diagram. It includes some species that may be found in pastures but are more characteristic of traditionally managed hay meadows, and analyses of modern pollen samples from sites with known land-use have shown a strong positive correlation between this pollen taxon and mowing (Gaillard et al. 1992). In the diagram from Värsjö Utmark

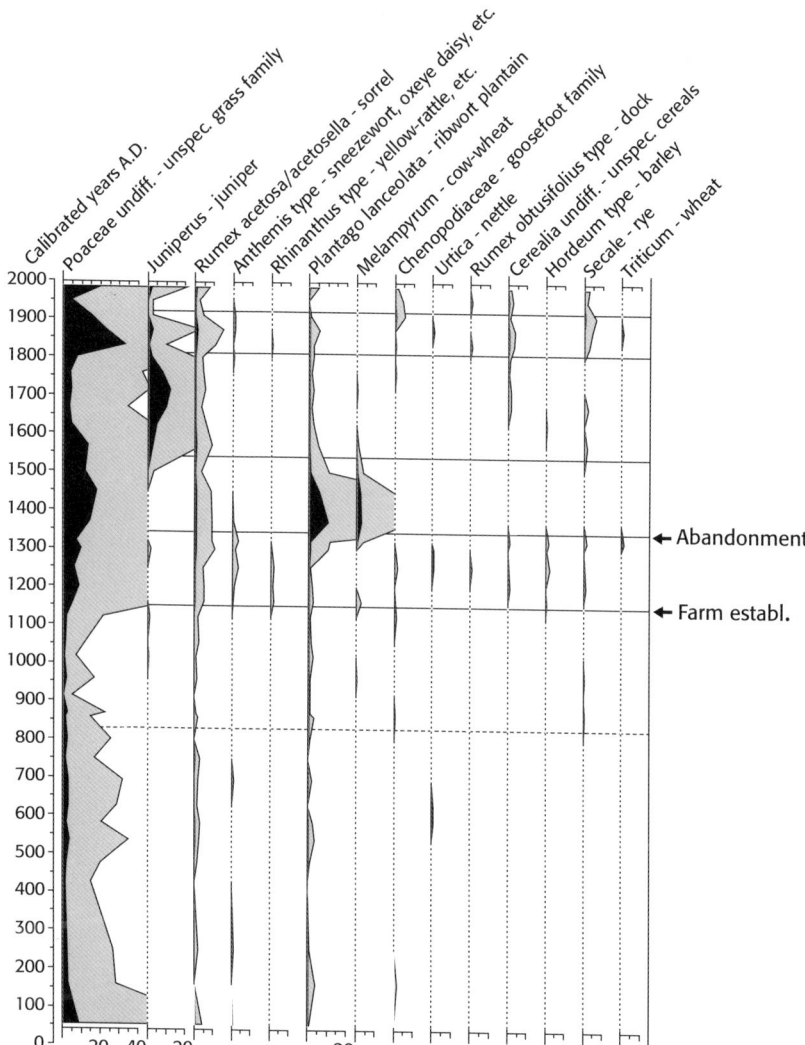

Fig. 23. Pollen diagram from Värsjö Utmark with selected taxa. The levels interpreted as reflecting early-medieval farm establishment and late-medieval abandonment are indicated. (For details see caption to fig. 9.)

Rhinanthus type shows a continuous record from the level of farm establishment until approximately the level interpreted as abandonment, and it is absent from the 14th century onwards in spite of the rich occurrence of other grassland taxa. The end of the *Rhinanthus* type record may tentatively be interpreted as reflecting a change in grassland use when hay meadows were transferred to pastures. Such an interpretation would fit in well with the interpretation above of a shift to a more extensive agricultural use of the area. It also fits in with the picture given by studies of written sources, for instance from the Vadstena convent, which shows that many abandoned farms in southern Sweden were used for grazing (Myrdal 2003).

To follow up the discussion on Grisavad above, it should be noted that the diagram from Värsjö Utmark does not show an increase in *Calluna* (heather) or any other signs of possible soil exhaustion before the abandonment in the 14th century. Taken together, the two sites give no support to the idea that farm abandonment was caused primarily by soil exhaustion, although there may of course have been local and regional differences.

According to the pollen analytical evidence presented so far, both Grisavad and Värsjö Utmark witnessed farm abandonment in the 14th century, and at both sites cultivation was re-introduced and settlement re-established in the 16th century. This late-medieval agricultural decline is however not found at the other two sites, Östra Ringarp and Bjärabygget. Östra Ringarp shows continuity in cereals and other agricultural indicators from the farm establishment in the 11th century, throughout the Middle Ages, and all the way up to the early 20th century. A decrease in cereal pollen percentages at a level dated to the 14th century may tentatively be interpreted as a temporary break or decline in cereal growing (figure 18), but it must, in any case, have been very short-lasting and there are no signs of farm abandonment.

Also at Bjärabygget there are no signs of a late-medieval decline. There are some changes in the pollen diagram at the level dated to the 14th century, such as a decrease in Poaceae undiff.

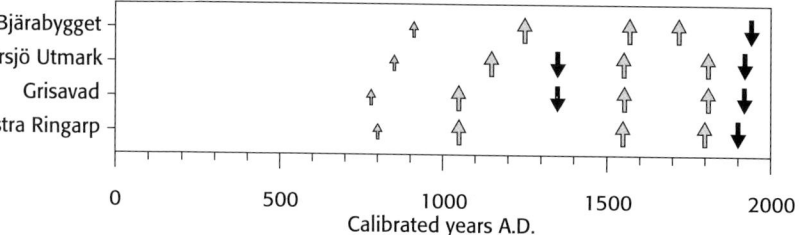

Fig. 24. Indications of agricultural expansion (grey arrows) and regression (black arrows) in the four pollen diagrams of the present study.

(grass) and an increase in *Betula* (birch), which may reflect overgrowing of some pastures, but the cereal record shows continuity indicating continuous cultivation. Also the radiocarbon-dated clearance cairns from the same site show continuity throughout the late Middle Ages (figure 17).

In summary, of the four medieval farms investigated in this study two were abandoned in the 14th century while the other two survived (figure 24). The abandoned farms were used extensively for grazing until the 16th century when new farms were established in more or less the same places and cultivation was re-introduced. In the meantime cultivation and settlement lingered on at the other two sites with no obvious signs of decline, but – which is interesting to note – also with no signs of expansion until the 16th century. Thus the overall impression on a regional scale is that the decline in the 14th century was followed by a period of stagnation that lasted for approximately two hundred years. During that period there was no capacity for agricultural expansion. With no doubt these changes, as documented in the pollen diagrams, reflect what is known as the late-medieval agrarian crisis (see next section).

The fact that some farms were abandoned and some were not raises the question of whether the character of the sites could explain this difference. Östra Ringarp is distinguished from the others in several ways, with early farm establishment, strip fields, fibre plant cultivation, and a large number of farmers mentioned in the 17th century sources. It is also the only one of the sites where there is still an existing farm today. In comparison to the others it thus has a rather central character and it is perhaps no surprise that it survived the late-medieval crisis. It is situated in the south-westernmost part of the investigation area at an altitude of 100 m above sea level, and has a combination of sandy till and sandy glaciofluvial deposits. The sampled peatland was a small lake during the Middle Ages.

The Grisavad site is situated only 1.6 km from Östra Ringarp and at the same altitude, and like Östra Ringarp it also shows early farm establishment. Grisavad has, however, a different environmental setting, dominated by glaciofluvial deposits and peatlands, which give it a rather poor character. This difference could possibly explain why the farm at Grisavad was abandoned in the 14th century while the farm at Östra Ringarp was not.

Moving to the northeast the farm at Värsjö Utmark was also abandoned in the 14th century. It was situated at an altitude of 120–130 m above sea level on a plateau of sandy till and peatlands. The relatively high altitude gives it a somewhat marginal character. However, the north-easternmost site, Bjärabygget, was also situated at 120–130 m above sea level but shows no farm abandonment in the 14th century. Furthermore Bjärabygget has a rather marginal character because of its situation on a peninsula of sandy till surrounded by peatland.

It is not easy to evaluate the different sites regarding their preconditions for agriculture, but based in particular on the Bjärabygget example we may conclude that there is no once and for all relationship between poor natural conditions and late-medieval farm abandonment. What is also interesting to note is that the investigated sites show no direct relationship between late farm establishment during the early-medieval expansion on the one hand and farm abandonment in the 14th century on the other. Otherwise one would perhaps expect that sites that were colonized late were so because of their marginal character and therefore would be abandoned first in times of crisis and decline. Perhaps there is such a relationship on a larger scale, but not among the investigated sites within the study area.

The Black Death and other possible causes behind the decline

The farm abandonment in northern Scania during the 14th century, which was inferred from pollen data and presented above, will now be put in a wider Swedish and European perspective. Based on thorough research by historians and others it is today a well-established fact that the early-medieval expansion was followed by a period of stagnation and population decline in the late 14th and early 15th centuries. This decline is known as the late-medieval crisis. It was a deep and far-reaching crisis characterized by decreased total production and it affected most parts of Europe and had an impact on all levels of society.

The most profound and striking evidence of the crisis is the large number of abandoned (or deserted) farms which indicate a population decline. These abandoned farms are frequently and specifically mentioned in early records, such as cadastral registers and other documents (e.g. Larsson 1970, Österberg 1977, Bååth 1983, Myrdal 2003). Some historians have raised doubts about what the term "abandoned farm" in the old records really stands for, i.e. if such a farm was uninhabited or only declared exempt from tax. However, in Sweden this seems to be a problem connected to registers from the 16th century onwards (Larsson 1972), but not to medieval ones in which the word abandoned (Sw: *öde*) really stands for uninhabited (Myrdal 2003). In addition to written records, sometimes the physical remains of the abandoned farms are also found in the field, in particular in forest or heathland outside modern cultivation districts. In Sweden such remains are found in different parts of the country and some of them have been investigated archaeologically (Gauffin 1981, Hansson et al. 2005, Åstran 2006). In the palaeoecological record, however, the late-medieval decline in Sweden has been rather invisible until now (see further discussion in the section *Linking crisis to the pollen record* below).

The most ambitious attempt to quantify the late-medieval farm abandonment in Sweden and its neighbouring countries was performed by the *Scandinavian Research Project on Deserted Farms and Villages* (Gissel et al. 1981). Based on a number of case studies in specific areas, the project presented an overview of the desertion frequency in the Nordic countries. According to that, more than 40% of the farms in Norway were abandoned, and also Denmark showed large-scale abandonment, while large parts of Sweden (including Scania) only showed a desertion frequency of less than 15% (Sandnes 1981). Unfortunately, this difference between the countries is however to a large degree only a consequence of the use of different methods (Österberg 1981). The Swedish part of the project used a method that gave erroneously low estimates of the real desertion frequency, and according to later research there was also large-scale abandonment in Sweden (Myrdal 2003). As an estimated average for southern and middle Sweden probably about 40–50% of the farms were abandoned, with regional differences ranging from more than 50% in forested regions to about 10% in central agriculture districts (Myrdal 2003). The estimated desertion frequency varies between authors, but there seems to be a general agreement that abandonment was most common in marginal areas that had been colonized relatively recently, i.e. during the early-medieval expansion (Larsson 1964, Myrdal 2003). The central areas were also hit by the crisis but as abandoned farms on good soils were soon taken over by new farmers the degree of long-term abandonment became rather low. Thus, the desertion resulted in a movement of people from marginal to central areas – a settlement concentration that can be seen not only in Sweden but also in several other European countries.

The farm abandonment obviously reflects population decline, but the desertion frequency is probably a little higher than that of population decline, because small farms in particular were abandoned. It is difficult to calculate the population decline, but in England, for example, it has been suggested that population numbers fell by as much as 50% or more between 1350 and 1450 (Hatcher 1977 in Morris 1989), and the general population decline in Europe has been tentatively estimated to be about 30% or more (Livi Bacci 2000). In Sweden there may also have been a major decline. In a recent study it has been suggested that the population drop between 1300 and the early 15th century in the area covered by present-day Sweden may in fact have been as high as 67% on average (Andersson Palm 2001).

In addition to the extent of the crisis, studies have also focused on its timing and character, and in particular on the possible causes behind the crisis. The most important single factor was certainly the Black Death, which will be discussed below, but there are also some other factors that have been discussed and which may have contributed to the decline. The reason why other contributing factors and explanations have been sought for is that old records indicate a negative trend starting back in the early 14th century, i.e. before the Black Death. Such indications are for example raising grain prices, but also specific evidence of famine, in particular from England (Kershaw 1973, Hybel 1988, Campbell 1995). Crop harvests in England and in other countries in northern Europe failed in 1315 and for several years to come, as a consequence of a series of unusually rainy summers, and, to make matters worse, failed harvests were followed by widespread murrains among sheep and cattle (Kershaw 1973). Apart from a series of rainy summers, the early 14th century has also been seen as the starting point for a long-term cooling trend after the Medieval Warm Period (Lamb 1995), although new data on mean temperatures for the northern hemisphere give a slightly different picture (Moberg et al. 2005; see figure 6).

It has also been suggested that the crisis should be seen as an ecological backlash after a period of excessively rapid expansion. One explanation put forward in England has been that the medieval expansion resulted in the colonizing of marginal areas with poor soils, whose productivity could not be sustained in the long term, and also that the demand for grain by the growing population resulted in an increase in arable land at the expense of pastures, which in the long run resulted in too little manure, soil exhaustion and falling yields (Postan 1972).

These different factors may have contributed to the decline but do not necessarily diminish the role of the Black Death. On the contrary, years of famine and other negative factors may have enhanced the devastating power of the plague by making people and society more susceptible and vulnerable. This argument is well established in English research. In Sweden, however, it has been argued that there are no or very few indications of decline in the early 14th century, and that the Black Death here should be regarded as the starting point for the crisis (Myrdal 2003, 2006). We should therefore now look more closely at the timing and the impact of the Black Death.

The Black Death was an epidemic disease that came from Asia and swept through Europe in 1347–1352. The general opinion is that it was bubonic plague, but other opinions have also been put forward (Cohn 2003). However, for the discussion here, it is not necessary to identify its medical character – only that it was a terrible epidemic with very high mortality. It reached Sweden and eastern medieval Denmark including Scania, and thus also the investigation area in northern Scania, in 1350. This was the first of a long series of deadly outbreaks that haunted the population for more than a century to come. The waves of the plague swept back and forth through Europe and hit different countries and areas at different times. According to compilations of several different sources it has been concluded that Sweden during the 14th and 15th centuries was hit by large-scale outbreaks in 1350, 1359–1360, 1368–1369, 1413, 1421–1422, 1439–1440, 1455, 1464–1465, and 1495, and in addition to these there were several smaller and spatially restricted outbreaks (Myrdal 2003). In England and several other countries the first outbreak – the Black Death by strict definition – was by far the most disastrous. In Sweden this was also the case, but here the second and the third outbreaks also showed devastating force. After the 14th century the plague continued to haunt the Swedish population throughout the 16th and 17th centuries. Some of these later outbreaks were disastrous enough, but the overall effect of the plague on population numbers and society was smaller than during the Late Middle Ages.

There was a major decrease in population due to the Black Death, and for a long time it was used as a time horizon in historical records – events were referred to as occurring before or after the Black Death. It remained in the common memory of society and is frequently mentioned in records from the 14th century all the way up to the present time. For example, the Swedish king Gustav Vasa wrote in a letter in 1555, i.e. two centuries after the Black Death, that it affected the people badly and that two thirds of the population died (Myrdal 1999a, Andersson Palm 2001). Gustav Vasa, together with several other writers, connected it with the remains of abandoned farms that could be found in forests and other marginal areas. A general opinion that prevailed for a long time was that the population was larger before the Black Death. Carl von Linné, for example, the famous botanist, wrote on his travels through northern Scania in 1749 that the ancient clearance cairns that he found in the forests were reminders of a large population before the Black Death (Linnæus 1751).

There is thus a long-held tradition that the Black Death resulted in a major drop in population in Sweden, but modern research by historians has resulted in different opinions about how big this drop was. It has been suggested that Sweden came off relatively well due to its small and scattered population (Nordberg 1995), but in the last few years new research and calculations have presented a completely different picture. According to this the Black Death hit Sweden very hard, at least its southern and middle parts, and the population decreased by 40–50% (Myrdal 2003), or possibly by as much as 60–70% (Andersson Palm 2001). In some marginal regions, population decline was probably even higher, due to migration to central areas where soils were more fertile. In a European perspective, population losses in southern and middle Sweden may have been slightly smaller than in the most severely affected countries of Europe, like Italy, France, England and Norway, but larger than in northern Sweden, Finland and much of Eastern Europe.

The Black Death was spread rapidly by fleas and rats and hit most of Denmark and southern and middle Sweden within a few months in the autumn of 1350. This first outbreak in 1350 definitely came as a shock, and there was no time for recovery before it was time for the next outbreak in 1359, another one in 1368, and so on. The force of the plague laid partly in its reoccurrence. As mentioned above, the outbreaks continued throughout the late 14th century, and also throughout the 15th, 16th, and 17th centuries. The last plague epidemic in Sweden was actually as late as in 1710–1713 (Persson 2001). It seems as if the force of the plague decreased during 15th and 16th centuries, but then increased again during the war-torn 17th century.

In Sweden the Black Death in 1350 was the starting point for the late-medieval crisis. It started a vicious circle, when disease and sudden population drop resulted in societal collapse, which in turn deepened the crisis, resulting in further population decline, etc. It became a deep and long-lasting societal crisis, which was reinforced by the continuously re-occurring epidemic outbreaks. The organization of society was difficult to uphold, leading to riots and civil war, and it was not until the 16th century that regression was definitely replaced by expansion and economic growth (Myrdal 2003, 2006).

To conclude, based on the latest research it seems clear that the Black Death was also a great disaster for Sweden and that it had far-reaching effects on society. A question that remains to be answered, however, is what effects it had on the landscape. The plague hit both central and marginal areas, but there are indications in historical records that it resulted in long-lasting farm abandonment mainly in forest regions and other marginal areas. Our study area in northern Scania is an example of such a marginal forest area, and the pollen diagrams from Grisavad and Värsjö Utmark show indications of abandonment in the

14th century. With a high degree of probability we may connect this abandonment to the Black Death. But the general impression from earlier pollen studies in southern Sweden has been that it is not possible to find any evidence of regression, which is surprising in the light of the historical evidence of a major drop in population and an average desertion frequency of almost 50%. This problem will be discussed in the next section.

Linking crisis to the pollen record: a new method for interpretation

In early pollen-analytical studies of the cultural landscape focus was on the distinction of periods of agricultural expansion and regression. Most influential was a paper by Björn Berglund in 1969, in which he used a series of regional pollen diagrams to identify four expansion stages within the time interval Neolithic–early Middle Ages (Berglund 1969). These expansion stages, which he regarded as synchronous over southern Sweden, and also partly over Denmark and southern Norway, had a great impact on archaeology and stimulated interdisciplinary studies of the landscape development (e.g. Gräslund 1979). Focus in Berglund's paper was on expansion stages, but between them were periods identified as stagnation or regression.

The chronologies of most of these early pollen diagrams were based on radiocarbon dates of bulk gyttja samples, which probably had errors due to the reservoir effect (Olsson 1986, 1991), or were based on long-distance correlation with other sites. Thus, the absolute chronologies were not very accurate, but on the other hand they may have been sufficient for the rather large-scale and long-term trends that the interpretations dealt with. Berglund illustrated the cultural landscape development with smooth hourglass-looking figures, which reflected a gradual and step-wise increase in landscape openness. These figures with ups and downs fitted in well with ideas on population growth and different levels of carrying capacity, which were discussed in the 1970s and 1980s (Boserup 1965, Welinder 1975, Emanuelsson 1988).

The expansion/regression terminology continued to play an important role for some time, not least in the large-scale and interdisciplinary Ystad Project in southern Scania, which in its general hypothesis stated that "the development of the agrarian landscape as seen in a long-term perspective is characterized by phases of expansion, consolidation, and regression" (Berglund 1991). The authors explained the expansions through different social and technological factors, and they concluded that the environment has responded to changes in society, but that it is difficult to find evidence of the opposite, i.e. that society has responded to independent environmental changes.

An impression from the Ystad Project is that it focused mainly on the expansion stages and that the intervening periods of regression were more difficult to explain. The regressions were characterized in the pollen diagrams as periods of extensive (in contrast to intensive) land-use and sometimes reforestation, in particular in marginal areas, and were also identified in the archaeological material as sparseness of settlement. The authors hesitated, however, to interpret the stagnation/regression stages as crises, and preferred to explain them as periods of settlement concentration, consolidation, and inner expansion. (It is worth noting that in this context they also found no unambiguous evidence of a regression during the Late Middle Ages.)

After the Ystad Project, scientific interest and project design within palaeoecology in Sweden has changed from regional to more local studies, and to some degree also from central agricultural districts to forested uplands. The development has partly been driven by problems formulated within forest ecology, where local studies may provide important insights into stand-scale dynamics, competition between tree species, etc. (e.g. Bradshaw 1988, Björkman 1996a). Also from a perspective of land-use and settlement history, the local pollen studies have a great advantage in providing detailed and spatially-precise information on local and short-term changes that are otherwise hidden in the smooth trends of regional pollen diagrams (e.g. Lindbladh 1998, Emaunelsson 2001).

If we now look more specifically at the late-medieval agrarian crisis, the general impression from the large number of pollen diagrams that have been published from southern Sweden so far is that this event is very hard to detect (Berglund et al. 2002). There may be several explanations for this. One of course is that there was no landscape change in connection with the crisis, but that seems unlikely in the light of the latest historical interpretations, pointing to large-scale farm abandonment and significant population drop (Andersson Palm 2001, Myrdal 2003). More plausibly the explanation lies in the character of the pollen diagrams and also in the scientific focus of the different studies and publications.

Many pollen diagrams from southern Sweden, in particular the early ones, have a regional character and thus provide a poor spatial resolution, as mentioned above. They also have a rather poor temporal resolution, which may have been sufficient for studies of long-term trends, but not for pinpointing a relatively short-term event as the late-medieval decline. Furthermore, their absolute chronologies are often problematic as they are based on radiocarbon dates of bulk gyttja samples. Due to the reservoir effect this dating material is expected to give erroneously old dates, with an average error of perhaps three or four hundred years but sometimes as much as one thousand years or more (Olsson 1986, 1991, Gaillard et al. 1996, Lagerås 1996b). In areas with calcareous soils the hard-water effect may result in even larger errors. Detailed and accurate absolute chronologies are not as important as long as interpretations and discussions deal with processes generally; however, they are the basis of a successful comparison with the historical record. Thus, in the 1970s and the 1980s, when there was some interest in detecting crises in the pollen record, most diagrams at hand were not very suitable for the task.

In the last two decades, the increasing awareness of the reservoir effect together with the development of the AMS technique has resulted in much better absolute chronologies. Bulk gyttja samples are now avoided, and the chronologies are based on AMS-dated terrestrial plant macrofossils, pollen concentrates or bulk peat samples. The more accurate chronologies and the increasing number of local pollen studies have improved our possibility of identifying a specific and relatively short-term phenomenon such as the late-medieval crisis. However, some of these local studies have not focused on land-use and settlement dynamics, but rather on forest ecology, and therefore the results have not been put into a historical and societal context. There is also a problem with local studies simply due to the fact that they are local. Having data that are spatially precise is in many ways a strength but it makes it difficult to draw any conclusions about the representativeness of the results and the overview is easily lost. A way to proceed is to combine several local studies into a regional picture – an approach of which the present study may be seen as an example. Only when put in a regional context, a vegetation change detected in a local diagram may with any confidence be interpreted in terms of a general process, such as, for instance, a late-medieval decline.

Encouraged by the results from Grisavad and Värsjö Utmark, which I interpreted as reflecting farm abandonment in the 14th century (see section *Local evidence of agricultural decline and abandonment* above), I have also looked for indications of regression in a number of other pollen diagrams. In an earlier study it was concluded that very few diagrams from southern Sweden show any indications of a late-medieval decline, but that compilation was based mainly on regional pollen diagrams from lakes with poor temporal resolution and more or less uncertain absolute chronologies (Berglund et al. 2002). The primary aim of that paper was to make a synthesis of the long-term cultural landscape development and not to pinpoint a short-term event such as the late-medieval crisis, which explains the choice of diagrams.

In the compilation that will be presented here, I have chosen only local pollen diagrams and among them only those that have a good chronology. I have also restricted the survey to marginal uplands, partly because most detailed local diagrams with a high resolution for the last millennium are from upland sites, but also partly because uplands are believed to have witnessed a higher desertion frequency than agricultural plains.

Altogether 20 diagrams (including the four from the present study) were regarded as suitable (figure 25, table 2). They all fulfil the demands of being local, covering the last two thousand years, having detailed and reliable absolute chronologies based on radiocarbon dates of terrestrial material, having a satisfactory temporal resolution (i.e. not having too few analysed levels per time unit), and, finally, having an uncomplicated depth/age relationship with no obvious hiatuses.

When it comes to identifying a relatively short-term event such as the late-medieval crisis the absolute chronology is definitely the weak point. Even if the radiocarbon dates are correct and based on reliable material, they always have a standard error, which for the late Middle Ages and some time before and after is usually approximately plus/minus 30–50 radiocarbon years (one standard deviation). When the radiocarbon dates are calibrated to calendar years, the error is also dependent on the shape of the calibration curve for that particular time interval. Thus, calibrated dates that fall within the 13th century have a one-sigma interval (total, not plus/minus) of about 50 years, for dates that fall within the early 14th century the corresponding interval is between 100–120 years, and for dates that fall within the late 14th century or the 15th century the interval is about 40 years. The relatively long calibrated intervals for the early 14th century depend on a wiggle on the calibration curve. In spite of this wiggle, the dating accuracy is however much better than for some other periods, such as, for instance the last three centuries before the present. A reasonable approximation is that changes during the late Middle Ages as reflected in pollen diagrams may be pinpointed to century level, but not more precisely than that. [In Ireland the mid-14th century level in pollen diagrams has been identified using the Öræfajökull tephra of A.D. 1368 (Hall 2003), but it is not known if this Icelandic tephra could be found also in Swedish bogs. The discovery and identification of late-Holocene tephras in south-Swedish bogs could be a future advance.]

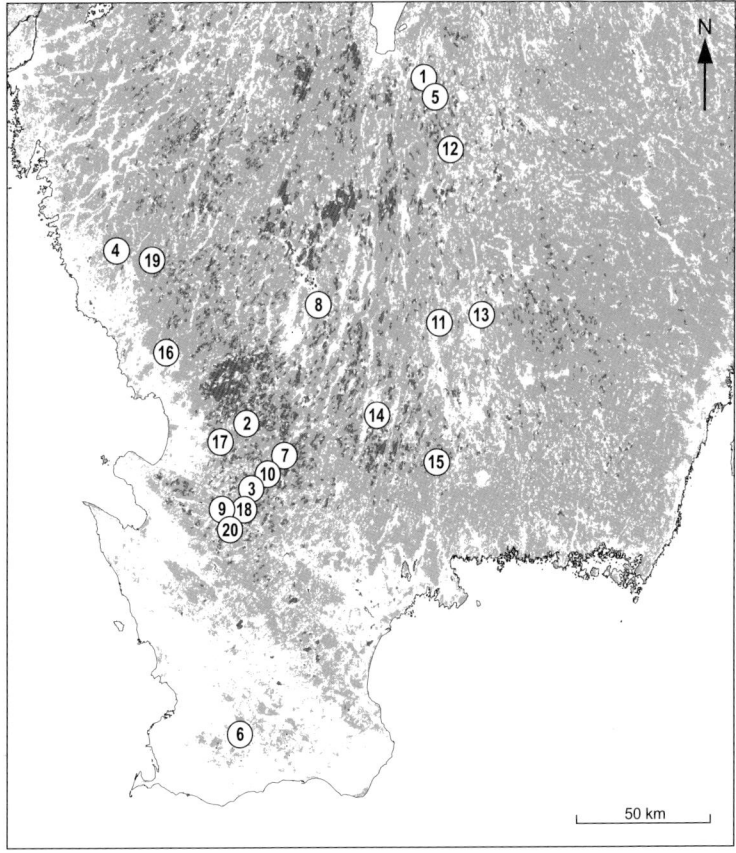

Fig. 25. Map of southern Sweden showing the pollen sites that are presented in table 2 and that are used for the compilation in fig. 26. Light shading indicates present-day forest cover and dark shading indicates peatlands. See table 2 for references.

The 20 diagrams have been searched for indications of expansions as well as regressions, and based on the available chronologies these indications have been dated at century level (figure 26). The interpretation of expansions and regressions is rather subjective and qualitative and based on the indicator-species approach (Behre 1981). The indications may be of different kinds at different levels and in different diagrams. A

Table 2. Sites used for the compilation in Figure 26.

Site	Expansion/regression	Reference
1. Avegöl 57°41'N, 14°29'E	Exp: 1st, 11th, 13th centuries Regr: 6th, 20th centuries	Lagerås 1996a, c
2. Baggabygget 56°35'N, 13°24'E	Exp: 12th, 16th centuries Regr: 20th century	Björkman 2005
3. Bjärabygget 56°21'N, 13°28'E	Exp: 10th, 13th, 16th, 18th centuries Regr: 20th century	this study
4. Bocksten A 57°07'N, 12°34'E	Exp: 8th, 18th centuries Regr: 19th century	Björkman 1996a, 1997a
5. Bråtamossen 57°40'N, 14°30'E	Exp: 13th, 19th centuries Regr: 6th century	Lagerås 1996a, Lagerås et al. 1995
6. Bökesjön 55°34'N, 13°26'E	Exp: 9th, 11th, 17th centuries Regr: 14th century	M.-J. Gaillard unpubl.
7. Exhult 56°29'N, 13°39'E	Exp: 9th, 12th, 17th centuries Reg: 20th century	Björkman & Ekström 2003
8. Flahult 56°58'N, 13°50'E	Exp: 11th, 16th centuries Regr: 15th, 20th centuries	Björkman 1996a, 1997b
9. Grisavad 56°17'N, 13°20'E	Exp: 8th, 11th, 16th, 19th centuries Regr: 14th, 20th centuries	this study
10. Köphult 56°25'N, 13°33'E	Exp: 13th, 16th, 17th, 19th centuries Reg: 20th century	Björkman 2003
11. Lekarydsdalen 56°55'N, 14°34'E	Exp: 4th, 9th centuries Regr: 6th, 14th centuries	Königsson 1989
12. Mattarp 57°29'N, 14°37'E	Exp: 8th, 13th, 16th, 19th centuries Regr: 10th, 20th centuries	Björkman 1996a, b
13. Rydholmskärret 56°57'N, 14°50'E	Exp: 2nd, 13th centuries Regr: 20th century	Ekström 2000, Lagerås 2002a
14. Råshult infield 56°37'N, 14°12'E	Exp: 4th, 12th, 16th, 19th centuries Regr: 14th, 20th centuries	Lindbladh & Bradshaw 1995, 1998, Lindbladh 1998
15. Siggaboda 56°28'N, 14°34'E	Exp: 12th century	Björkman 1996a, Björk- man & Bradshaw 1996
16. Trälhultet 56°48'N, 12°54'E	Exp: 12th, 19th centuries Regr: 20th century	Björkman 2000a
17. Uddared 56°31'N, 13°15'E	Exp: 12th, 17th centuries Regr: 15th, 19th centuries	Björkman 2000b
18. Värsjö Utmark 56°19'N, 13°26'E	Exp: 9th, 12th, 16th, 19th centuries Regr: 14th, 20th centuries	this study
19. Yttra Berg 57°05'N, 12°49'E	Exp: 12th, 16th centuries Regr: 14th, 19th centuries	Sköld 2006
20. Östra Ringarp 56°16'N, 13°19'E	Exp: 9th, 11th, 16th, 19th centuries Regr: 20th century	this study

change interpreted as agrarian expansion may for example be an increase in Cerealia (cereals), an increase in Poaceae undiff. (grass), or an increase in *Calluna* (heather). Usually a combination of observations is used, so that an increase in *Calluna* is interpreted as expansion only if Cerealia does not decrease at the

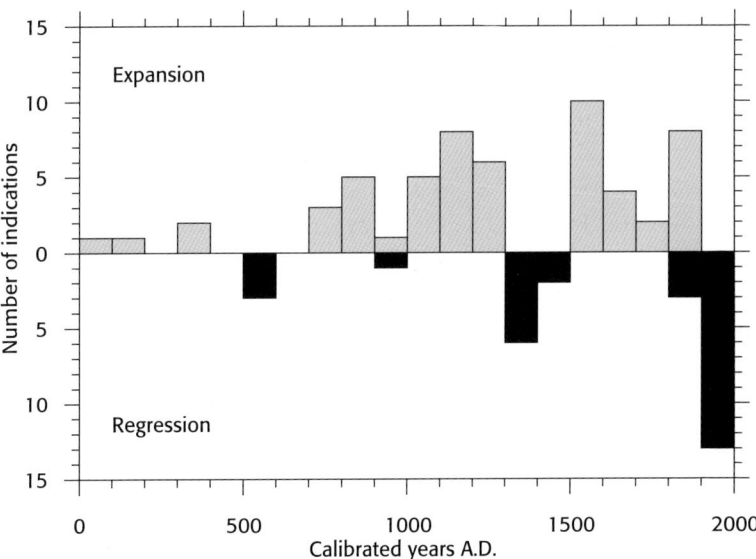

Fig. 26. *Indications of agricultural expansion (grey bars) and regression (black bars) in 20 high-quality pollen diagrams from upland areas in southern Sweden (see fig. 25 and table 2). Bars show number of indications per century.*

same level, which in that case could reflect a change from intensive to extensive land-use. Changes interpreted as agrarian regression may for instance be a break in the cereal record, a decrease in grass and herb pollen, an increase in *Betula* (birch), etc. It is important to point out that it is not possible to use one type of indication for all times and for all sites. Poaceae undiff., for instance, may be a good grazing indicator at one site but may reflect natural bog vegetation at another (e.g. purple moor-grass *Molinia caerulea*). The same is true for *Calluna*, which may reflect grazed heathland in one diagram but natural bog vegetation in another. This problem can be sorted out by comparison with graphs for other taxa, and by checking the documented peat lithology. However, it exemplifies why interpretations of expansions and regressions preferably have to be qualitative and consider several different aspects, rather than relying on a once-for-all parameter, such as, for instance the ups and downs of the NAP curve (i.e. the sum curve for all non-tree pollen).

The number of indications per century, of expansions on the one hand and regressions on the other, is presented in the bar chart in figure 26. In summary, there are 56 indications of expansion and 28 of regression, which may be regarded as a reflection of a general increase in human impact during the last two millennia. But what is more interesting in this context is the temporal distribution of the indications. Examining the bar chart from left to right, i.e. in chronological order, reveals that the first indications of agrarian expansion are dated to the first centuries A.D. (N.B. that these are the first only in a time-perspective of the last two millennia and that there may have been expansions during earlier periods as well). The first regressions are dated to the 6th century, i.e. the Migration Period, when three of the sites show indications of farm abandonment (Lagerås 1996a). After that follows a period characterized by expansion in all the 20 diagrams. From this very pronounced expansion period, which lasted from the 8th century to the 13th century, there are altogether 28 indications of expansion and only one of regression. Although the Middle Ages does not start until A.D. 1050 according to the Swedish time scale, this entire expansion period may be referred to as the early-medieval expansion in a wider north-European perspective.

The shift from early-medieval expansion to regression comes out surprisingly clearly in the bar chart. From the 14th and 15th centuries there are indications of regression at eight sites. Two of the sites are Grisavad and Värsjö Utmark of the present study, and the other six are Bökesjön, Flahult, Lekarydsdalen, Råshult infield, Uddared, and Yttra Berg (for references see table 2). For three of them, the regression had been identified and connected to the late-medieval decline in the original papers (Königsson 1989, Lindbladh & Bradshaw 1998, Sköld 2006). But what is equally interesting is that not a single one of all the 20 sites shows any indication of agrarian expansion of any kind during the 14th or the 15th centuries. So, in addition to the signs of regression at some sites, all the sites together contribute to the picture of a stagnation period that lasted for approximately two hundred years. Thus, the compilation shows that the late-medieval agrarian crisis is readily visible in the local pollen diagrams from upland areas in southern Sweden, not necessarily as indications of regression but as a lack of indications of expansion. Such an interpretation could not be made based on a single site, but only on a compilation such as this one.

After the regression and stagnation period there was a new period of expansion, which lasted from the 16th century to the 19th century. From this period there are 24 indications of expansion. Finally, the 20th century is in the bar chart strongly characterized by regression (to some degree starting as early as the 19th century), which reflects the large-scale landscape change when crofts and other small farms in the uplands were abandoned and at the same time modern silviculture was introduced and marginal agricultural land was transferred to forest. In contrast to the late-medieval decline this regression does not reflect a general crisis, but rather the polarization of the south-Swedish cultural landscape into open agriculture districts (not included in this compilation) and forest regions (included).

To sum up the conclusions and discussion in this section, it is possible to identify the late-medieval agrarian crisis in pollen diagrams from southern Sweden – not only in the new ones from northern Scania but also in other upland sites. Some local diagrams (eight out of twenty; i.e. 40%) show indications of overgrowing or other signs of agrarian regression during the 14th and 15th centuries, while all the studied diagrams give a unanimous picture of stagnation. Signs of regression are perhaps the ones that are most easily interpreted in terms of an agrarian crisis, but stagnation may also be seen as a sign of a crisis, in particular in this case when it follows a 600-year period of overwhelming expansion.

This new picture of the late-medieval crisis inferred from pollen diagrams was partly the result of a new method for interpretation. Qualitative indications for expansion and regression were compiled in a bar chart, which in turn enabled a quantitative evaluation of expansion and regression phases in a south-Swedish perspective.

16TH CENTURY RE-EXPANSION

·········

Local evidence of agricultural expansion

After the late-medieval agrarian crisis followed expansion. Slowly and step by step the negative trend, which had started in the early or mid 14th century, was broken and both population numbers and agricultural production started to increase again. The timing of this change differed from country to country and also from region to region, and due to the complex character of the development it is not possible in any case to put an exact date on the end of the crisis. In a general European perspective, the 16th century in particular was a period of expansion, although a positive trend in some countries started as early as the late 15th century. The same is true for Sweden and Denmark, where the re-occupation of abandoned farms and the establishment of new ones in some regions started as early as the mid 15th century, but a more general rise in population is evident from the early 16th century onwards (Larsson 1972, Myrdal 1999a).

The different mechanisms behind the expansion are difficult to sort out, but important factors may be found in the technological development – such as, for instance, the development of larger and more efficient iron shares for ards and ploughs – as well as in changes in society (Myrdal 1999b). An important factor in Sweden and Denmark was that the state grew strong, increased its possession of land, and developed a more advanced administration. In order to stimulate expansion and population growth the state offered exemption from tax for new settlers in marginal areas, usually for a period of ten to twenty years after farm establishment. In Sweden the state claimed possession of all outlands, which then could be offered to new settlers without obstruction from other farmers. The reasons for the state to encourage agrarian expansion in this way was first of all to increase its income in taxes in the long run, but also to populate uninhabited marginal areas along the national borders.

In addition to agricultural expansion and a stronger state, the 16th century was characterized by increased division of labour, well-developed markets and far-reaching trade, and also by technical progress within non-agricultural sectors (Söderberg & Myrdal 2002). Hence, there was an overall expansion in society and different sectors affected each other in a complex way. For instance, increased agricultural productivity and efficient markets enabled non-agricultural sectors to expand, which resulted in technological development, increased iron production, etc., which in turn resulted in increased agricultural productivity. (Evidence of local iron production will be discussed in a separate chapter below.)

The agricultural expansion during the 16th century in Sweden was most clearly reflected in the large-scale establishment of new farms in marginal areas. Even though Denmark has a more fragmentary source material from this period, it is clear that it also witnessed agricultural expansion. In particular the second half of the 16th century was a time when many new single farms were established in the forested uplands of northern Scania and in other marginal areas of the Danish kingdom (Skansjö 1997b).

As in most other periods there was a mutual relationship between agricultural production and population numbers. Colonization of marginal areas and other expansion of cultivated land were dependent on increasing population, and, *vice versa*, a strong population rise would not have been possible without an increase in the total agricultural production. It seems as if Sweden, in a European perspective, had an unusually vigorous rise in population, possibly due to the room for expansion offered by sparsely settled forests and uplands. For the area of present-day Sweden it has been estimated that population numbers between 1500 and 1600 rose from 450,000 to 750,000 (Andersson Palm 2001).

Turning to our investigation area in the uplands of northern Scania, the pollen diagrams enable us to look for local expressions of the 16th century expansion and observe how it is reflected in vegetation and landscape development. In particular the two sites Grisavad and Värsjö Utmark are interesting in this respect, as they had witnessed farm abandonment in the 14th century.

At Grisavad cultivation ceased in the 14th century and cultural land was partly overgrown by shrubs and trees. The local farm was abandoned but some pastures were kept open by grazing animals, probably belonging to one or several farms situated some distance away. In the pollen diagram, the cereal record starts again at a level dated to the 16th century, which reflects the re-introduction of cultivation and the re-establishment of a local farm (figures 27, 29). The cereal frequencies continue to rise at levels representing the 16th and most of the 17th centuries, and then remain high until the uppermost part of the diagram representing the 20th century. During the first phase of the expansion cultivation is represented by *Secale* (rye), *Hordeum* type (barley), and *Triticum* (wheat), while *Avena* (oat), *Fagopyrum esculentum* (buckwheat), and *Linum usitatissimum* (flax) appear somewhat further up. In addition to cultivated taxa, the expansion is reflected in increasing values also for some other pollen types, such as *Rumex acetosa/acetosella* (sorrel) and Poaceae undiff. (grass), although a more sharp increase in Poaceae undiff. waits until the late 17th century.

The re-introduction of cultivation during the 16th century is not only reflected in the cereal record and in other open-land taxa, but also in a sharp decline in *Fagus* (beech). *Fagus* showed increasing values after the 14th century abandonment, reflecting how beech forest expanded on abandoned land. Thus, the succeeding decline in *Fagus* during the 16th and early 17th centuries not only shows that beech forest was now cleared to give way to fields and other cultural land, but also that the cleared areas were the same ones that had once been open during the Middle Ages but had overgrown in the meantime.

Similar to Grisavad, the Värsjö Utmark site shows evidence of ceasing cultivation in the 14th century, indicating farm abandonment. What differs is that the diagram from Värsjö Utmark

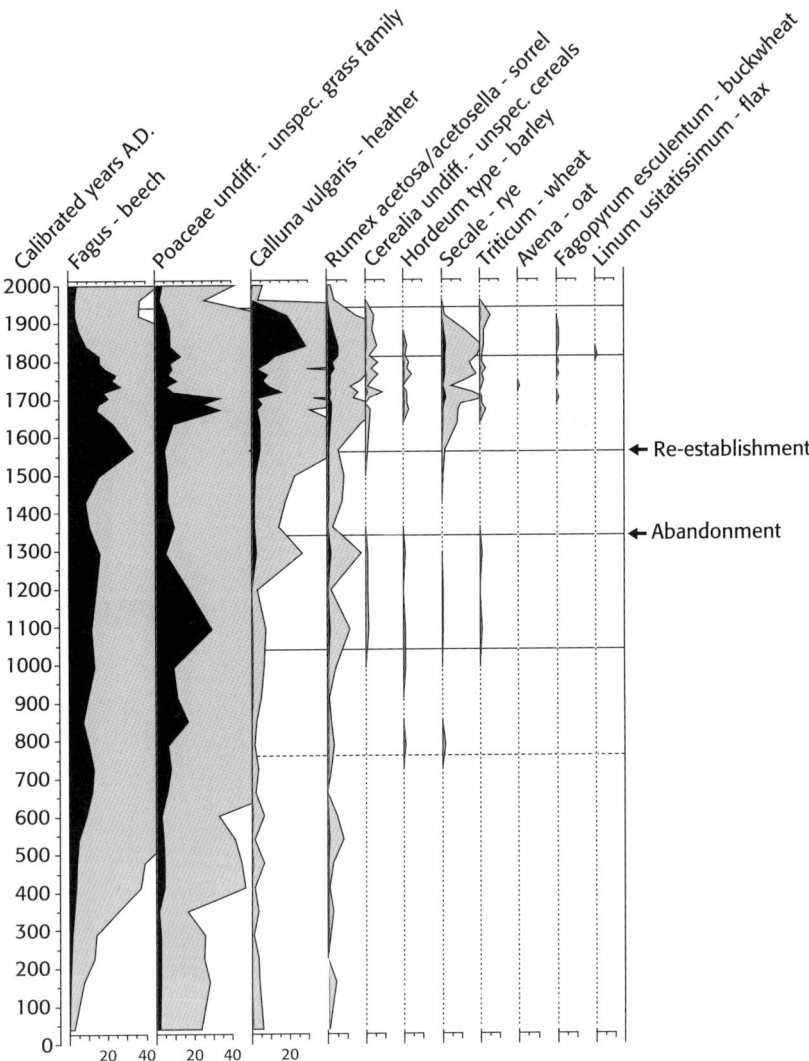

Fig. 27. Pollen diagram from Grisavad with selected taxa. The levels interpreted as reflecting late-medieval abandonment and 16th century re-establishment are indicated. (For details see caption to fig. 9.)

shows no sign of overgrowing. The area around the site, including abandoned fields, was kept open by grazing, and there
are indications that also hay meadows were transformed into
pastures. However, during the later part of the abandonment
period – i.e. approximately during the later part of the 15th century – increasing *Betula* (birch) values may indicate some overgrowing due to a decreased grazing pressure.

The re-introduction of cultivation at Värsjö Utmark is indicated by the appearance of cereal pollen at a level dated to the
16th century (figures 28, 29). At the same level there is also a
decrease in *Betula*, which may reflect clearing of overgrown
pastures. However, the picture from Värsjö Utmark is not as
straightforward as the one from Grisavad. At the same level
where cereal pollen reappears and *Betula* declines there are
also an increase in *Juniperus* (juniper) percentages and a decrease in Poaceae undiff. (grass). It seems as if some grasslands
were overgrown by juniper shrubs at the same time as arable
fields were established and cultivation re-introduced. A possible explanation is that the establishment of a farm in the area
was accompanied by a structural change of pastures leading to
decreased grazing in some areas. Another possible explanation
is that juniper shrubs were actually wanted, as they contributed with raw material for fencing. It is also known from written
sources that the strong tradition of basket-making in this part
of northern Scania used mainly juniper (Bringéus 1983).

At the other two sites, Östra Ringarp and Bjärabygget,
there was no farm abandonment in the 14th century and thus
no re-establishment should be looked for. However, it can be
noted that in particular the diagram from Östra Ringarp, but
also to a lesser degree the diagram from Bjärabygget, show increasing cereal pollen percentages at levels representing the 16th
and the 17th centuries (figure 29). This may reflect new farms
being established or just an expansion of arable by existing
farms. It is also worth noting that at Östra Ringarp the increase in cereal is not accompanied by any significant rise in
Poaceae undiff. (grass) or other grassland indicators (figure 9).
Even though it is difficult to quantify the extent of arable based

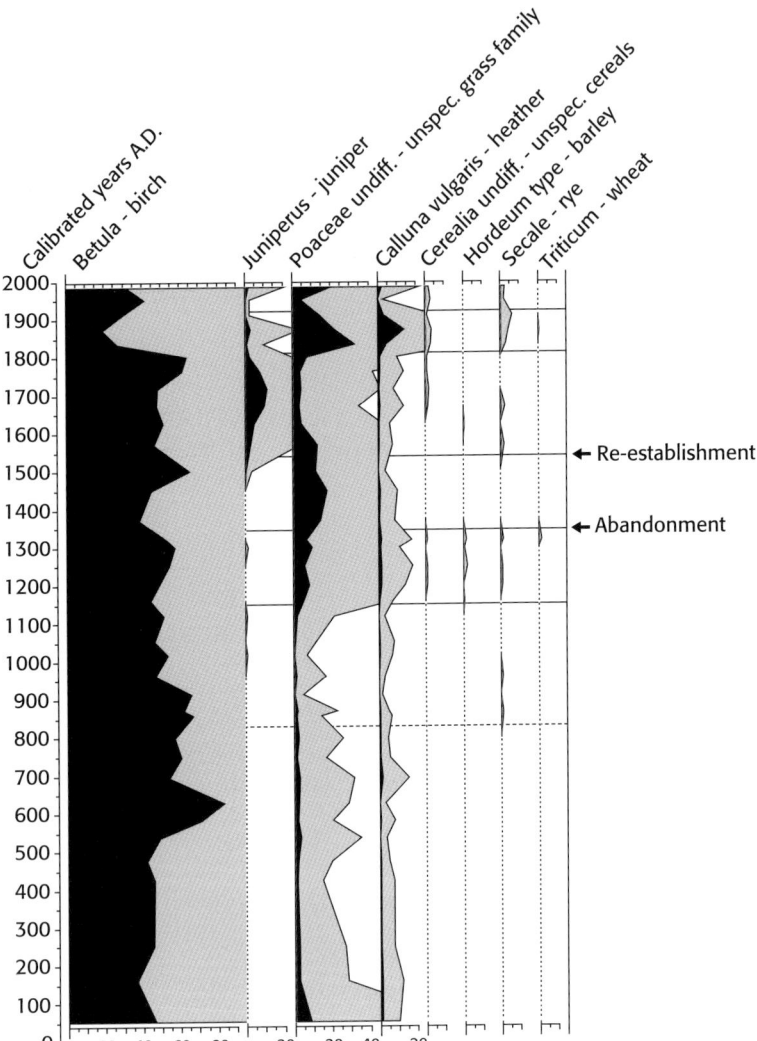

Fig. 28. Pollen diagram from Värsjö Utmark with selected taxa. The levels interpreted as reflecting late-medieval abandonment and 16ᵗʰ century re-establishment are indicated. (For details see caption to fig. 9.)

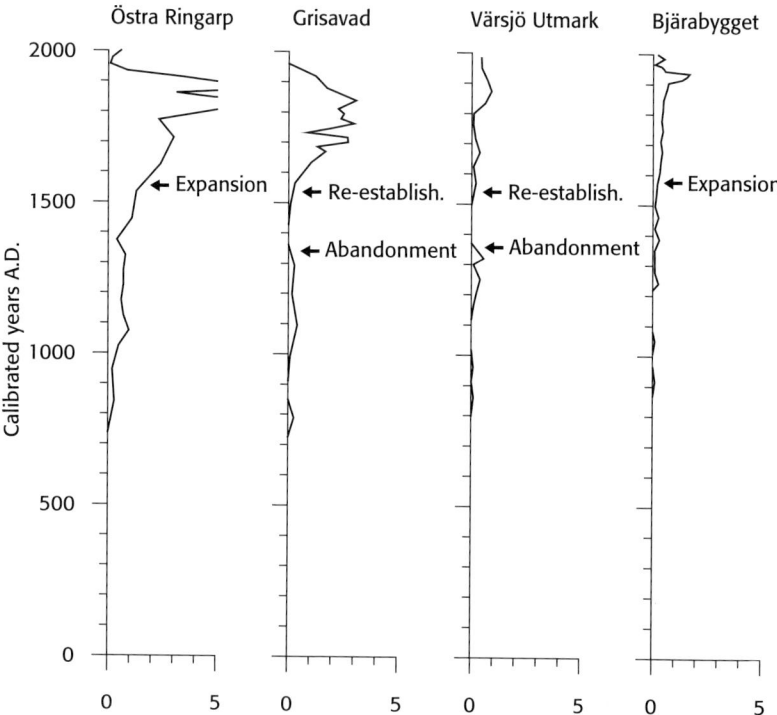

Fig. 29. *Summary graphs for all Cerealia pollen [i.e. Cerealia undiff. (unspec. cereals), Hordeum type (barley), Secale (rye), Triticum (wheat), and Avena (oat)] from each site. The graphs show percentages of total pollen. Levels interpreted as reflecting late-medieval abandonment and 16th century re-establishment/expansion are indicated.*

on pollen frequencies, this change may tentatively be interpreted as an increase of arable in relation to pastures, and hence an intensification of the land-use.

To sum up, all the four pollen diagrams in the present study reflect agricultural expansion during the 16th and 17th centuries, in particular an expansion of arable but at some sites also of pastures and probably hay meadows. According to the diagrams this was the first agricultural expansion in the investigation area for three hundred years, i.e. since the medieval establishment of a farm at Bjärabygget in the 13th century, which was the last expression of the early-medieval expansion. Also if we look at the compilation of a larger number of pollen diagrams from southern Sweden presented in figure 26, it is evident that in particular the 16th century stands out as a period of agricultural expansion. We may conclude that the 16th century expansion, which is well documented in historical sources, is clearly reflected in south-Swedish pollen diagrams.

The agriculture practised during the 16th and 17th centuries seems to have been rather similar to the one practised before the late-medieval agrarian crisis, at least as far as can be judged from the pollen data. We find more or less the same combination of cereals – i.e. rye, barley, wheat – during both periods, and it was not until later, during the 18th century, that new crops such as oat and buckwheat were introduced. Also the pastures of the 16th and 17th centuries seem to have been similar to the medieval ones in that they were dominated by grasses. Heather started to expand later, during the 18th century and in particular during the 19th century, when there was more extensive heathland.

As mentioned in the beginning of this section there may have been a complicated combination of positive factors that contributed to the 16th century expansion. Such factors were for instance the development of more efficient iron tools for agriculture, exemptions from tax and other central initiatives to encourage colonization of marginal areas. Most important were also the increased division of labour and the improved conditions for trade. However, there were also some factors which one would expect to have had a more negative effect, such as climatic deterioration and war. These will be looked at in the next two sections.

The Little Ice Age

Climate change has frequently been used to explain history, and climatic deterioration in particular has been interpreted as a decisive factor behind crises in society, from small-scale abandonment to the collapse of entire cultures and dynasties. The character and quality of the palaeoclimatological data that have been used for comparison with historical and archaeological records have been shifting, but generally the quality has improved greatly during the last few decades due to extensive research. Awareness of the human-induced greenhouse effect and the global warming of the last century – which leads to worries about future climate change – have resulted in a great demand for new information on past climate and its natural variability. Today there is an overwhelming amount of palaeoclimatological information available, from proxy data and site-specific interpretations to global circulation models and computer-based reconstructions and predictions of past and future scenarios.

There are many different kinds of palaeoclimatological information but it may be divided into three main categories (Lamb 1995): instrumental records, historical sources, and analytical data from stratigraphies and other natural archives. Instrumental records, i.e. first-hand records of meteorological measurements of temperature, precipitation, etc., are only available for the last three centuries and in most areas for a much shorter period than that. Historical records reach further back in time, and represent a great diversity of different sources. They range from intentioned weather diaries and mention of extreme weather situations, to indirect indications such as grain-price records, building-repair records, etc. Written sources dominate, but also paintings, for instance, may show the former distribution of glaciers in the Alps, winter ice on the Thames, etc. The third category of palaeoclimatological information is the wide range of data provided by analyses and measurements of different physical and palaeoecological parameters. The source material is fossil records such as ocean sediments, peat deposits, glacier ice cores, tree rings, end moraines, etc., and examples of parameters may be stable isotope relationships, pollen frequencies, peat humification, etc. This type of data reaches thousands of years back in time, and is the most important for long-term reconstruction of past climatic change.

In southern Sweden palaeoclimatological research has mainly focused on the late Pleistocene and the early Holocene, while details of the climatic development of the last millennium have not gained as much attention as in England and some other countries (cf. Lamb 1995). There are probably local historical records from southern Sweden that could be used, but no accessible compilations have been published and there is no data available for our investigation area in northern Scania.

With the lack of detailed local studies, the best basis for a discussion on the climatic development is the temperature curve for the northern hemisphere recently published by Moberg et al. (2005) (figure 30). It is based on a combination of high-resolution tree-ring data and low-resolution proxies from lake and ocean sediments (and for the last two centuries also instrumental data), and reveals both year-to-year variation and long-term trends in northern hemisphere mean temperatures. The curve represents two thousand years but focus here will be on the last millennium. As evident from the curve, the Early Middle Ages were characterized by a favourable climate – a period known as the Medieval Warm Period. In particular the period 950–1150 was warm, with peaks around 1000 and 1100. After the warm period followed a period of colder climate usually referred to as the Little Ice Age. In a wide sense the Little Ice Age started with a cooling trend in the 13th century and ended with the increasing temperatures of the second half of the 19th century. However, the medieval cooling trend was slow and step-wise and it was not until the 16th and 17th centuries that climatic bottom values were reached.

The Little Ice Age was a global phenomenon (Grove 2001), which has traditionally been explained as a climatic response to a combination of natural factors. Of these the most important ones were reduced solar activity and enhanced volcanic activity (Luterbacher 2001). There is no doubt that this cold period had a strong effect on people and society, exemplified in numerous ways by the historical record (Lamb 1995, Fagan 2000): crop growing failed, vineyards closed down, glaciers advanced, fishing failed due to ice-blocking of harbours, Arctic sea-ice expanded southward (leading to the surprising appearance of Eskimos in their kayaks in Scotland!), etc. However, recently a

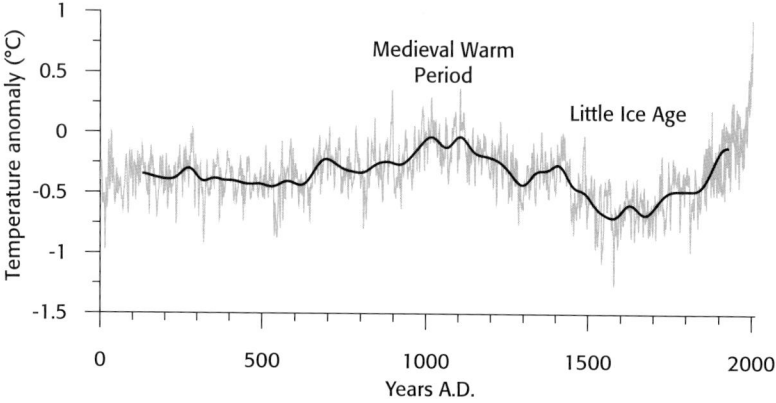

Fig. 30. Mean temperature curve for the last 2000 years in the Northern Hemisphere (Moberg et al. 2005). (For details see caption to fig. 6.)

rather thought-provoking hypothesis has been put forward, suggesting that the Little Ice Age was not only a natural phenomenon that affected human society, but that this cold period may in fact have been at least to some degree induced by humans (Ruddiman 2005). Analyses of ice cores have shown that the Little Ice Age, like most cold periods, was characterized by relatively low concentrations of carbon dioxide (CO_2) in the atmosphere. The concentration of atmospheric CO_2 reflects climate but since CO_2 is a greenhouse gas it not only reflects climate but also affects it. According to Ruddiman the low concentrations during the Little Ice Age cannot be explained by natural factors alone, but may be a reflection of large-scale reforestation on abandoned agricultural land. Population decline and farm abandonment due to the Black Death and other factors led to the expansion of forests, and since forests are CO_2 consumers the forest expansion led to decreased CO_2 concentrations in the atmosphere. (Today we have the same process in reverse when large-scale deforestation in the tropics leads to increased CO_2 concentrations and rising temperatures: the 'greenhouse effect'.) Although controversial, this hypothesis highlights the complexity of the human–nature relationship and shows how society, agriculture, vegetation, and climate are related to each other even though the causal relationships may be difficult to sort out.

Leaving the question of the possible causes of the Little Ice Age, we can conclude that it was a profound climatic event with a global distribution, and even though we do not know the details, we can be sure that this cold period was also effective in the investigation area. For most of southern Sweden and Denmark the Little Ice Age probably meant colder and drier winters than normal and also colder springs and wetter summers (Luterbacher 2001). In 1658 the Swedish army, led by king Charles X Gustavus, crossed the ice on Store Bælt Sound and reached Copenhagen in a surprising attack, after which Denmark was forced to give up Scania and several other provinces. This event may be seen as a dramatic testimony of the cold winters of the Little Ice Age in southern Scandinavia, and also perhaps as a fascinating example of how climate may affect history. It is, however, not a typical example. There may be some more occasions like this – when extreme climatic situations affected the decisions of kings and the outcome of war – but the most important way in which climate has generally affected society is through its influence on agriculture, and thus on the livelihood of common people.

Depending on natural preconditions, together with social and technological factors, different areas show different vulnerability to climatic change. In particular in environmentally marginal areas, climatic changes have been decisive for the possibility to grow crops as well as for other food production. The limiting climatic factors vary between different parts of the world, but in northern Europe agriculture is mainly limited by low temperature, which leads to short vegetation periods and risk of frost, and high precipitation during the growing season. In particular at high latitudes and/or high altitudes, where cereal growing was barely possible even during periods of relatively good climate, a climatic deterioration may have resulted in failed harvests, and a series of failed harvests may in turn have led to famine and farm abandonment.

It is difficult to say if the uplands of northern Scania were close to a climatic boundary in a way that made them sensitive to a period of climatic deterioration such as the Little Ice Age. The reason for this difficulty is that we do not know the magnitude of the cooling in the area, and also that we do not know the exact climatic tolerance of the crops and the agricultural system that was practised. In other upland areas it has been convincingly shown that the cold climate of the Little Ice Age resulted in farm abandonment, for instance in Scotland (Parry 1981). In comparison to the Scottish Highlands, the altitudes of the investigation area are not very high, ranging from 80 to 160 m above sea level, and abandonment due to climatic deterioration should perhaps not be expected. What can be concluded, however, is that agriculture must have been more problematic in the years of the Little Ice Age, in particular during the 16ᵗʰ and 17ᵗʰ centuries, than during the warm Early Middle Ages when the area was colonized.

What may be surprising is that the pollen diagrams show agrarian expansion during the 16ᵗʰ century when the temperatures of the Little Ice Age reached bottom values. So, in contrast to the early-medieval colonization and agrarian expansion, which coincided with an unusually favourable climate, the 16ᵗʰ century expansion coincided roughly with the coldest period of the last two millennia. This may be seen as a paradox and it shows that there is no simple and direct relationship between climate and agricultural development, and that any attempts to interpret the causal relationships also have to involve other factors.

One way to discuss the impact of climate on agriculture is in the terms of risk (cf. Carter & Parry 1984). The risk of harvest failure is to a high degree climate-dependent, but what level of risk is acceptable depends on other factors, which are mainly social and economic. A mean frequency of harvest failure of, for instance, one year in five may be acceptable in one society or within one economic system but not within another. In the investigation area, the decreasing temperatures that mark the transition from the Medieval Warm Period to the Little Ice Age probably increased the risk of harvest failure, in particular in increasing the probability of late frost in the spring or too much rain in the late summer leaving the grains to rot in the fields. The fact that there was agrarian expansion during the cold 16th century may tentatively be interpreted as a reflection of an increased acceptance of a higher risk within agriculture. This acceptance may in turn be due to a broader economic basis for the farmer's livelihood and for society as a whole. As mentioned in the previous section, the 16th century was characterized by increased division of labour, with a growing proportion of the population working outside the primary agrarian sector (Söderberg & Myrdal 2002). In the investigation area in northern Scania, iron production in particular was of great importance, providing alternative incomes to the traditional farming. This will be discussed in the chapter *Iron and Charcoal Production*.

War

Similar to the poor climate of the Little Ice Age, which was discussed above, one would expect war also to have had a negative effect on the agricultural development of the 16th and 17th centuries. Sweden and Denmark were at war with several other countries during this period, for instance in the large and devastating Thirty Years' War (1618–1648), but in particular they were at war with each other. A long period of more or less constant battle between Denmark and Sweden started with the onset of the Nordic Seven Years' War in 1563 and continued throughout the 17th century and ended with a last act of war in 1710. The Scanian population suffered greatly as Scania was the main arena for much of the fighting between the two countries (Skansjö 1997b). Periods of intense battle in Scania during this period were 1563–1570, 1611–1613, 1657–1658, 1676–1679, and 1709–1710. The complicated details of the politics and battles will not be dealt with here, but one of the major outcomes was that Denmark lost Scania and several other provinces to Sweden. Formally, Scania became a Swedish province in a treaty in 1658, and Danish attempts to recover it in 1676–1679 and 1709–1710 failed.

The war was cruel and cost many lives. The cruelty may be exemplified by the exceptionally bloody battle at Lund in 1676, where approximately ten thousand soldiers died in one day. But the civilian population also suffered badly. Taxes were raised to finance troops and rearmament. Farms and households were plundered as armies, both Swedish and Danish, again and again marched back and forth across the province. In the wake of war followed starvation and epidemics, not least the plague, which had its last outbreak in Scania in 1710–1713. As a consequence of the intense and long-lasting wars of the 17th century, the agricultural expansion and the population rise, which had characterized much of the 16th century, slowed down and in the most war-torn parts of northern Europe, such as northern Poland and Germany, population rise was turned into decline. For Scania it has been suggested that there was no significant decline in population, but the expansion stopped and during the late 17th century population numbers remained relatively constant (Myrdal 1999a).

The upland areas in northern Scania were borderland between the two nations and were therefore particularly hard struck. Since the Seven Years' War, or even earlier, it was a well-established strategy on both sides to plunder and burn the borderland in order to prevent alien troops from establishing themselves (Österberg 1971). After the conquest of Scania, Swedish troops were frequently attacked and ambushed by local guerrillas, and they retaliated with merciless punishments and executions of war captives and civilians. For instance, by order of the king in 1678, all farms in the parish of Örkened in northeastern Scania, which was regarded as particularly guerrilla-infested, were burnt down and the fields were devastated (Skansjö 1997b).

If we now turn to the landscape development of the investigation area, as reflected in the pollen diagrams, surprisingly we find no clear evidence of abandonment during the 17th century. On the contrary, the diagrams show constant pollen values for cereals at two sites (Värsjö Utmark and Bjärabygget) and increasing values at the other two (Östra Ringarp and Grisavad) (figure 29). Hence, there is no indication of abandonment of arable fields. The picture is more complicated when it comes to pastures. At Östra Ringarp not only cereals but also Poaceae undiff. (grasses) and most other open-land indicators show very constant values, indicating stable conditions. At Grisavad there are some short-term changes, with strong peaks of Poaceae undiff., *Myrica* (bog-myrtle), *Alnus* (alder), and *Pinus* (pine), at levels dated to the late 17th and early 18th centuries. These peaks are difficult to interpret but may reflect some kind of disturbance or short-lasting interruptions. At Värsjö Utmark there are increasing percentages of *Juniperus* (juniper) reflecting overgrowing of some pastures. Finally at Bjärabygget, a decrease in Poaceae undiff. and an increase in *Betula* (birch) is noted at a level dated to the mid-17th century. It may reflect some overgrowing of pastures, but as early as the early 18th century the birch woodland was cleared and heathland was established.

To sum up, there is no evidence of abandonment of arable fields. There seems to have been some disturbance in the grazing regime, which tentatively may be interpreted as a decrease in the number of grazing animals, but there was no significant reforestation. The possible overgrowing of pastures was brief, and there is no sign whatsoever of permanent abandonment. Due to the long calibration intervals of radiocarbon dates from the 17th century and later, there may be some uncertainties in the time-scales of the pollen diagrams during the period discussed here. But even if the chronologies should be corrected a century back or forth in time, there would be no evidence of abandonment.

The picture given by the four pollen diagrams from northern Scania is supported by the compilation of twenty South-Swedish pollen diagrams presented in figure 26. According to the compilation, agricultural expansion during the 17th century was not as vigorous as during the 16th century, but what is important to note is that none of the twenty diagrams show any sign of regression. The conclusion must be that there was no major abandonment of agricultural land in the uplands of southern Sweden during the 17th century, in spite of the war and the well-documented difficulties for the population. Of course, there may very well have been farm abandonment and local reforestation in some areas – for instance in the parish of Örkened (see above) – but in general it seems as if agricultural land-use went on as before and that disturbances and overgrowing were only temporary.

When it comes to the investigation area in northern Scania, the possibility cannot be excluded that the war also had some positive effects on the local economy. The war meant increasing demands for timber for military shipbuilding and for iron for cannons, cannonballs, etc. As the area was rich in both timber and bog ore it may to some degree have profited from the war. In particular iron production is evident from the large number of iron production sites that have been recorded in the area and of which many have been dated to the 16th and 17th centuries. Iron production and its importance for the local society and landscape are discussed in the next chapter.

IRON AND CHARCOAL PRODUCTION

·········

Evidence of iron production

The production of iron has a long history, which in Sweden can be traced back to the Late Bronze Age, c. 700 BC. Soon after its introduction iron replaced bronze as the main metal for making implements and tools. It has several advantages, for instance it may be forged to strong, slender blades, but its greatest advantage was probably its availability. In contrast to bronze, which had to be imported from Continental Europe, iron could be produced locally from bog ore found in peatlands in many parts of Sweden. Iron soon came to be used for agricultural tools, first of all for sickles, but from the Roman Iron Age onwards also for scythes (Myrdal 1982, Pedersen & Widgren 1998). The sickle was used for harvesting cereals and leaf fodder, and the scythe for cutting hay. Even though the early sickles did not differ much from their bronze predecessors and the early scythes were short-bladed and rather inefficient, these cutting tools may have been a contributing factor behind the agrarian expansion witnessed in many areas as early as the Roman Iron Age (e.g. Lagerås 2000b). The sickle and the scythe were further developed throughout the Iron Age and the Middle Ages, with longer blades and new types of shafting, and at the same time iron also started to be used for the production and improvement of other agricultural implements. Tips and shares of iron were put on the wooden spades and ards (Myrdal 1999a), which made them stronger and enabled more efficient tilling, not least in the stony tills of upland areas. The heavy iron axe was also an important improvement which in a very direct way facilitated agrarian expansion and colonization of woodlands by making woodcutting and deforestation easier.

Thus, a general relationship may be seen between iron production and the medieval expansion, as iron implements came

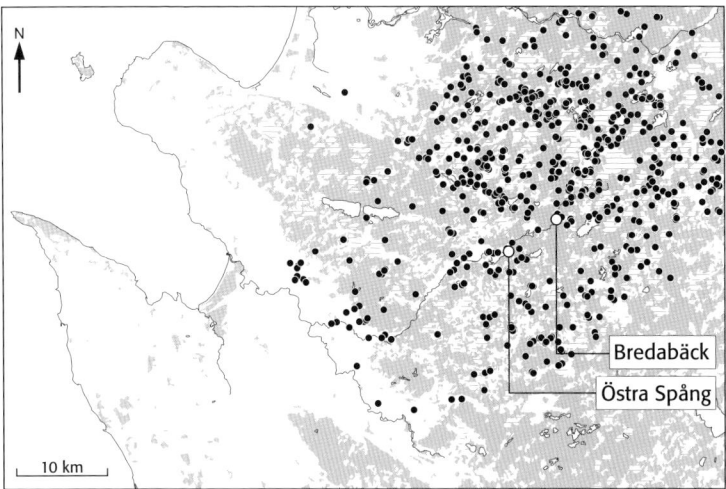

Fig. 31. Map of northern Scania and neighbouring provinces showing ancient bloomery iron production sites. Based on FMIS: the National Heritage Board's digital register of ancient monuments. The two iron production sites Bredabäck and Östra Spång, which were excavated within the present project, are indicated. Shading indicates present-day forest cover and hatching indicates peatlands.

to be widely used and enhanced the possibilities for creating and maintaining fields, pastures and meadows. This is true for most of the parts of northern Europe that witnessed a medieval expansion. In the investigation area, however, apart from this general relationship, iron production may also have had a more direct impact on the vegetation and landscape development, as large-scale iron production was carried out within this actual area. The investigation area is situated in a part of the South-Swedish Uplands that is unusually rich in remains of ancient bloomery iron making (figure 31). Slag heaps, furnaces, and other remains are found in such large numbers that there is no doubt that this upland area was once an important centre for production.

The slag heaps and other remains of local iron production in the area have attracted attention from archaeologists and historians, and some sites have been excavated and dated (Wedberg 1981, Englund 1995, Rubensson 2000, Ödman 2001). Two sites were also excavated as part of the present project (figure 31). One of them was a production site situated close to a peatland and a small brook named Bredabäck, three kilometres northeast of Åsljunga (and less than one kilometre south of the pollen-analysed site Värsjö Utmark). Among the investigated features were the remains of furnaces, anvil stones, slag, and charcoal, and the activity period was radiocarbon-dated to the 13th–14th centuries. The other site excavated within the project was situated by the river Pinnån, at Östra Spång, two kilometres east of Örkelljunga. Here, the remains of a very large and well-built furnace were discovered and investigated, together with a smaller furnace, slag, and the bottom layer of a charcoal kiln. The activity on this site was radiocarbon-dated to the late 15th–early 17th centuries. The results of these excavations are presented in more detail in a separate book from the project (Strömberg in prep.).

By not only using the dates from these two excavations, but also compiling all available radiocarbon dates from iron production sites in the area, we get a fairly good overview of the distribution of this activity in time (figure 32). Iron production in the area started in the late 12th or early 13th century and it continued throughout the rest of the Middle Ages and early Modern Time, until the mid-17th century when it ceased. During this four to five century-long period of iron production in the area, there seem to have been no major interruptions, although shorter breaks (shorter than a century) cannot be ruled out. However, based on the frequency of dates, two phases may be distinguished: an early one which lasted from the late 12th or early 13th century until the 15th century, and a late more intense one which lasted from the late 15th until the early 17th century. A change is also noted regarding the character and environmental setting of the sites. The ones from the early phase are relatively small and situated typically in close connection to peatlands, while during the late phase we see a larger diversity of sites. The peatland settings were still used, but now larger

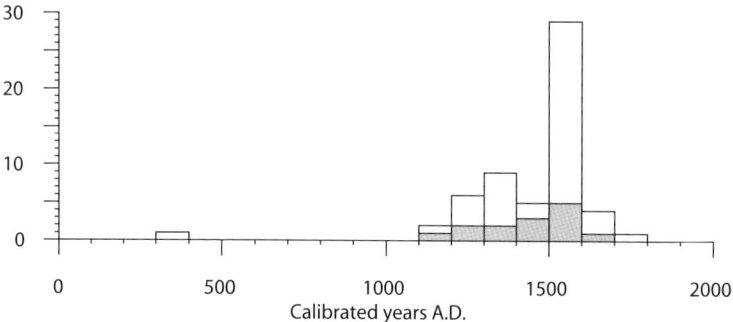

Fig. 32. Compilation of radiocarbon dates from bloomery iron production sites in northern Scania, southern Halland, and southwestern Småland. Grey bars show dates from Bredabäck and Östra Spång (see fig. 31), while hollowed bars show dates from other projects (Wedberg 1981, Englund 1995, Rubensson 2000, Ödman 2001). The bars show the number of dates within each 100-year interval, and are based on the mid-points of the calibrated one-sigma intervals. The bar chart is based on 57 radiocarbon dates.

production sites were also established by lakeshores, and also by rivers and brooks. The latter reflects the introduction of waterpower to run bellows and hammers, while the setting on lakeshores indicates that now iron ore also taken from lake sediments contributed to the production. The second phase of iron production – which was more intense and had a more industrial character than the early phase – reflects the technological progress and the increased division of labour, which were two important factors behind the 16th century expansion (Söderberg & Myrdal 2002).

To sum up, iron production in the area started in the late 12th or early 13th century, became large-scale and more industrialized from the late 15th century onwards, and ceased in the 17th century. Initially in the project, it was hypothesized that the suitable conditions for iron production – i.e. the richness of both bog ore and wood – may have been an important driving force behind the colonization. This may be partly true, as the early-medieval expansion was still going on during the 12th century and even later, exemplified in the present study by farm establishment during the 12th and 13th centuries at Värsjö Utmark and Bjärabygget. However, the pollen diagrams from Östra Ringarp and Grisavad, situated in the southwestern part of the investigation area, show farm establishment as early as the 11th century. It should also be noted that farm establishment at all sites was preceded by temporary cultivation from the 9th century onwards. Thus, it seems as if the onset of colonization and agrarian expansion pre-dates the earliest evidence of iron production in the same area. We may therefore conclude that the first phase was mainly agrarian, and that the idea of a colonization process driven by the demands for iron has no empirical support when the chronological relationships are examined in more detail.

After its introduction, iron production soon became an important part of the local economy, and it met not only demands from within the area but also supplied distant markets. The explanation as to why this area became a centre for iron production is of course partly to be found in its natural resources, but also in its location in relation to political borders. During the Middle Ages and up to the late 17th century (formerly until 1658) the province of Scania (and for most of the time also Halland and Blekinge) belonged to Denmark. The central parts of the Danish kingdom, i.e. the densely populated agricultural plains of southern Scania and Zealand, were short of woods and peatlands, and their need for iron and iron products had to be fulfilled through import from surrounding uplands and other marginal areas. The uplands in northern Scania, in the borderland to Sweden, was one of very few such areas within Denmark that had suitable conditions for iron making. Its large-scale production reflects the demands for iron within Denmark – from agriculture, from cities, and from the military. The increased production in the second phase may partly be explained by increased demands from the military due to frequent wars during the 16th and 17th centuries (see previous section).

The end of the production in the 17th century may also to some degree be explained by the political situation, as it coincided with the final Swedish conquest of Scania. In the new situation of being part of Sweden – which in contrast to Denmark was very rich in woods and peatlands – the uplands of northern Scania were no longer unique resource areas. However, the most important factor behind the decline was probably that bloomery iron making, based on bog and lake ore, was becoming old-fashioned and lost out in competition with the new technique of melting bedrock ore in blast furnaces. This new technique was introduced at some sites in Sweden as early as the Middle Ages, but it was in strong expansion particularly during the 17th centuries, when high quality Swedish iron was exported all over northern Europe (Heckscher 1957). Probably the production in northern Scania would have been driven out of business by Swedish iron production even without the conquest, but perhaps not as suddenly as now seems to have been the case. In the shadow of this new technique, however, small-scale bloomery iron making continued, both in the investigation area and in other parts of Sweden, all the way up to the 19th century, but it was of minor economic importance.

Impact on woodlands

From the perspective of palaeoecology an interesting question is what impact the large-scale iron production may have had on the landscape, but this question is more difficult to answer than one would perhaps expect. What makes it complicated is that the period of iron production was also a time of agricultural impact on the landscape, which makes it difficult to distinguish the environmental effects of one activity from the other. In earlier studies in other areas, attempts to use pollen analysis to evaluate the impact of iron production and to separate it from agriculture have also highlighted this problem (cf. Königsson & Qvarfort 1988, Karlsson 2000). Still, despite the difficulties, some conclusions may be drawn.

The major impact of iron production on the landscape was not caused directly by the iron-making itself, but indirectly through its great demand for fuel. Bloomery furnaces were fired with wood and charcoal, and – based on the richness of furnaces, slag heaps, etc. in the area – a reasonable suggestion is that fuel collection must have affected the woodlands in northern Scania in some ways, possibly leading to deforestation. However, as mentioned in the previous section, the earliest farm establishment and the onset of major deforestation in the area started in the 11th century, which predates the earliest iron production by one or a few centuries. Therefore the initial deforestation reflected in the pollen diagrams cannot be interpreted as an effect of iron production and its need for fuel. Even if some small-scale production of iron and charcoal may also have occurred in this early phase, the deforestation and the creation of a semi-open landscape – as witnessed in the pollen diagrams – must be interpreted as a reflection of agricultural expansion.

Somewhat later, from the 13th century onwards, we have reliable dates from iron production sites both in the investigation area and in the region as a whole, and we thus have an empirical basis for suggesting that fuel collection may have contributed to deforestation during that phase. This may have been the case particularly at places such as Värsjö Utmark and Bjärabygget, where pollen analyses show deforestation not during the 11th century but during the 12th and 13th centuries. An interesting example is the investigated iron production site at Bredabäck, which was dated to the 13th–14th centuries and situated less than one kilometre from the Värsjö Utmark site. Obviously this production site was in use at the same time as the nearby farm at Värsjö Utmark, which was established in the 12th century and abandoned in the 14th century. The deforestation reflected in the pollen diagram from Värsjö Utmark may not only have had an agricultural purpose, but may also to some degree have been the result of fuel collection for iron production at Bredabäck or at some other nearby site. Of course, one explanation does not necessarily exclude the other, as tree felling may have produced fuel for iron production and new agricultural land at the same time. However, the pollen diagrams from Värsjö Utmark and other sites convincingly show that after deforestation cleared areas were managed and kept open by continuous cultivation, grazing and probably hay mowing, and therefore, still, the main driving force behind the early-medieval deforestation and landscape change was probably the agrarian expansion as such and the need for new agricultural land.

Some time after the early-medieval phase of colonization and farm establishment, which ended in the 13th or possibly early 14th century, iron production in the area increased in importance. Starting probably as early as the late 15th century and no later than the 16th century, the iron production activity was intensified, leading to larger and more industrialized production sites. The production was at its peak during the 16th and the early 17th centuries. During this period the pollen diagrams show some indications of deforestation that to some degree may be attributed to fuel collection for iron and charcoal production. However, this was also a period of agricultural expansion, clearly evidenced by an increase in cereal pollen in all the four pollen diagrams. In particular the 16th century was a period of farm establishment and agricultural expansion all over the investigation area. Therefore, not even this phase of intensified iron production – when it became large-scale and reached a more industrial character – can be convincingly identified and distinguished from agricultural impact in the pollen diagrams. It is plausible, however, that the iron production during this phase resulted in some deforestation although this is difficult to prove.

Even if iron production together with agriculture led to deforestation, there seems to have been a great deal of woodland remaining in the area even in the late 17th century, i.e. immediately after the period of the most intense iron production. This conclusion may be drawn from the pollen diagrams, but it is also evident from an old map dated to 1684 (figure 43a). The map shows roads, farms, churches, etc., as well as the distribution of woodland. With two different symbols it even separates deciduous woodland from coniferous woodland. According to the map there was a great deal of woodland in northern Scania at the time, and in the investigation area the woodland was mainly deciduous. This picture fits in well with the one given by the pollen diagrams. In the same way as the map the pollen diagrams show a strong dominance of deciduous tree taxa over coniferous taxa in the 17th century. According to the diagrams, coniferous trees started to expand somewhat later, first with pine (*Pinus*) in the 18th century and later with spruce (*Picea*) in the 20th century. And like the map, the pollen diagrams also suggest a relatively large degree of forest cover at the time. Even if it is difficult to quantify forest cover in absolute numbers based on pollen data, the relatively high percentages of tree taxa (the average frequency of tree pollen in the four diagrams at this level is approx. 80%) suggests a forest cover of between 40 and 80% (cf. Sugita et al. 1999). A calculation based on the map from 1684 gives a forest cover of 45%. Even though the results of these calculations are tentative, they strongly indicate that it was not a shortage of woodland that brought iron production to an end.

Initially in the project, it was noted that the investigation area was not only rich in remains of ancient iron production, but also that it had been relatively rich in heathland (see the chapter *Man-made Heathland: Birth and Decline* below). These heathlands are now gone, due to modern forestation, but they were relatively widespread until the early 20th century. It was hypothesized that fuel collection and deforestation in connection with iron production contributed to the initial establishment of these heathlands, and that they were subsequently managed and kept open by grazing. Based on the results of pollen analysis and the dating of iron production sites the hypothesis could however be rejected. The pollen analysis showed that even if there may have been some restricted heathland early on, the main heathland expansion was during the 18th and in particular during the 19th century, i.e. well after the iron production ceased in the 17th century.

Leaving the question of whether and to what degree iron production may have led to deforestation and increased landscape openness, the next question is whether it may have affected the composition of the forest. A source material that can be used in this context is macroscopic charcoal data. We do not know if un-charred wood was used in the bloomery furnaces as well, but from the archaeological excavations within the project we do know that charcoal was used. Evidence of both the production and the storing of charcoal was found in close connection to furnaces, slag layers, etc. at the two investigated iron production sites (Strömberg in prep.). Such charcoal can with certainty be connected with the iron production. In addition, charcoal production sites were also found at other excavation sites within the project. Even if these lack a close spatial relationship to known iron production sites, they may still, at least some of them, have produced charcoal to be used in the iron production.

The macroscopic charcoal data used here are charcoal pieces that have been sampled during excavation, identified to tree genera and then radiocarbon-dated. The results are presented in two tables, one for iron production sites (table 3) and one for charcoal production sites (table 4). The results in table 3 show that all charcoal from the iron production sites originates from deciduous trees. Birch (*Betula*) is strongly dominating (>80%) but many other tree taxa are also represented: alder (*Alnus*), beech (*Fagus*), oak (*Quercus*), hornbeam (*Carpinus*), cherry (*Prunus*), hazel (*Corylus*), maple (*Acer*), lime (*Tilia*), alder buckthorn (*Frangula*), ash (*Fraxinus*), and aspen (*Populus*). The dates span the entire iron production period from the 12th century to the early 17th century, and throughout this period no significant changes in the charcoal composition are observed.

When the charcoal composition at the iron production sites is compared to the forest composition according to pollen data, we find there is fairly good agreement. As mentioned above, the pollen diagrams show that the woodlands during the period of iron production were dominated by deciduous trees and that the major expansion of coniferous taxa – i.e. pine (*Pinus*) and spruce (*Picea*) – started later (figure 33). The dominance of deciduous trees in the charcoal data thus seems to simply reflect the woodland situation at the time. Another agreement

Table 3. *Charcoal composition of samples from iron production sites that were investigated and radiocarbon dated within the project. The sampled contexts are slag layers and bloomery furnaces, and they are presented in the table in chronological order (V24:1=Östra Spång, E4:31=Bredabäck).*

Site/ID	Lab.no	C14 BP	Cal 2s (max –min) A.D.	Charcoal taxa (no. of ident. fragments)
V24:1/117	GrN-28687	280 ± 20	1520–1670	Betula 345, Alnus 22, Fagus 12, Carpnus 9,
	GrN-28333	315 ± 15	1510–1650	Prunus 5, Tilia 2, Acer 1
V24:1/2127	GrN-28330	315 ± 25	1480–1650	Betula 26, Alnus 5, Tilia 2
V24:1/8921	GrN-28684	320 ± 20	1490–1650	Betula 70, Alnus 7, Quercus 5, Prunus 4
V24:1/105	GrN-28685	300 ± 25	1490–1660	Betula 104, Alnus 5, Fagus 3, Frangula 2, Populus 1,
	GrN-28686	375 ± 20	1440–1630	Carpinus 1
	GrN-28688	365 ± 20	1450–1630	
	GrN-28689	440 ± 25	1420–1485	
V24:1/2387	GrN-28683	380 ± 20	1440–1630	Betula 1
E4:31/5282	Ua-26874	830 ± 40	1060–1290	Betula 12, Alnus 1, Fraxinus 1
E4:31/3407	Ua-26872	855 ± 40	1040–1280	Betula 36, Corylus 9, Quercus 7, Acer 4

Table 4. *Charcoal composition of samples from the 12 different charcoal-production sites that were investigated and radiocarbon dated within the project. The sites are presented in chronological order. Note the dominance of deciduous trees in the early ones and of Pinus in the later (E4:34a=Värsjö Utmark).*

Site/ID	Lab.no	C14 BP	Cal 2s (max –min) A.D.	Charcoal taxa (no. of ident. fragments)
E4:55/100001	Ua-26879	0	Rec.	Pinus 15
LV:10/100001	Ua-26880	0	Rec.	Pinus 30
E4:35/100001	Ua-26875	75 ± 40	1680–1960	Pinus 9
E4:44/100001	Ua-26878	80 ± 40	1670–1960	Pinus 11
E4:34a/728	Ua-26828	105 ± 30	1670–1960	Pinus 135, Betula 4
	Ua-26833	175 ± 30	1650–1960	
E4:34a/686	Ua-26832	195 ± 30	1640–1950	Pinus 111
	Ua-26831	205 ± 30	1640–1950	
LV:10/100008	Ua-26881	200 ± 40	1640–1960	Pinus 30
E4:17/261	Ua-26862	295 ± 35	1480–1670	Betula 51, Alnus 12
	Ua-26753	325 ± 30	1480–1650	
V24:4/588	Ua-26864	305 ± 35	1480–1660	Fagus 27
V24:1/5507	GrN-28332	375 ± 20	1440–1630	Betula 58, Alnus 19, Prunus 4, Fagus 2
E4:34a/2053	Ua-26948	780 ± 45	1160–1300	Pinus 38, Fagus 30
	Ua-26949	790 ± 45	1160–1300	

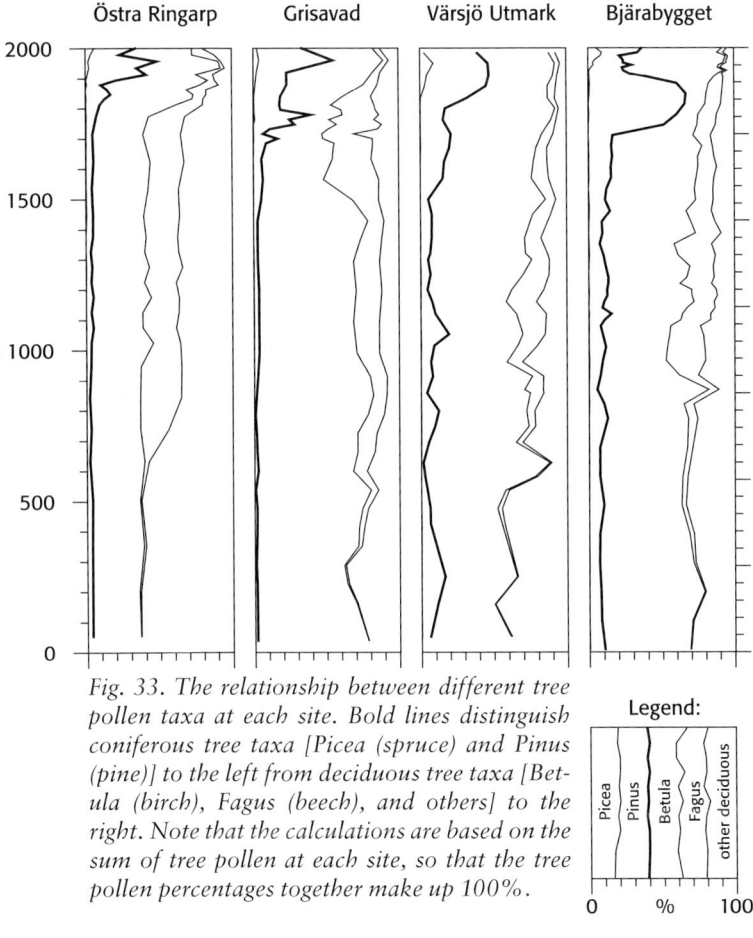

Fig. 33. The relationship between different tree pollen taxa at each site. Bold lines distinguish coniferous tree taxa [Picea (spruce) and Pinus (pine)] to the left from deciduous tree taxa [Betula (birch), Fagus (beech), and others] to the right. Note that the calculations are based on the sum of tree pollen at each site, so that the tree pollen percentages together make up 100%.

between charcoal and pollen data is the strong dominance of birch (Betula). In layers from this period it makes up on average 40–50% of the total pollen sum and 50–60% of the tree pollen sum, which makes it by far the most dominant tree pollen taxon. However, birch is a strong pollen producer, and it is certainly over-represented in pollen data in comparison to its actual abundance in the woodlands. Its true abundance is difficult to estimate, but based on earlier studies a pollen frequency for Betula of 50–60% of the tree pollen may tentatively be interpreted as indicating that birch did not make up more than 20% of the tree cover (cf. Andersen 1970, Sugita et al. 1999, Broström et al. 2004). In comparison to that figure, the fact that as much as 80% of the identified charcoal at the iron production sites come from birch indicates that it was a preferred tree when it came to charcoal production. It may reflect that birch was not regarded as a valuable tree for other purposes, in the way in which for instance oak (*Quercus*) and beech (*Fagus*) were important mast trees. As birch is a light-demanding early-succession tree, it may also reflect that fuel collection for charcoal production did not primarily take place in dense forests but rather in more open, grazed woodlands and other parts of the cultural landscape.

If we now turn to the results from the charcoal production sites presented in table 4 we get a somewhat different picture. The radiocarbon dates from these sites cover a larger time span than the data from iron production sites, stretching from the 12th century more or less to the present. Thus, we have evidence of charcoal production from the time when iron production was being carried out in the area, as well as charcoal production that was carried out considerably later. Based on the chronological relationship, the early sites may tentatively be interpreted as serving the bloomery furnaces with charcoal, but the later ones must have had another purpose. (It should be noted that the later ones are more difficult to date in detail than the early ones, as a plateau on the calibration curve results in long calibration intervals for the period *c.* 1650–present. There is no doubt, however, that they are from the late 17th century or later, i.e. after the period of iron production.)

If we look at the charcoal composition at these sites, we find an interesting pattern: the early sites, similarly to the iron production sites, show a strong dominance of deciduous trees, while the later ones are strongly dominated by pine (*Pinus*). Obviously, deciduous trees were preferred for coaling during the period of iron production, but later, for some reason, the coaling shifted focus to pine. It may to some degree reflect how pine became a more common tree in the landscape in general (figure 33), but also that there was a change in the spatial pattern for fuel collection. Most of the pine woodlands were to be found on poor sandy soils and in particular on peatlands, while in more central parts of the agrarian landscape there were still deciduous trees such as birch (*Betula*), oak (*Quercus*), and beech (*Fagus*). When we look at the spatial distribution of the charcoal production sites, we also find that most of the early ones are located in the southern, more densely settled part of the investigation area, while the later ones dominated by pine charcoal are situated in the northern part where there is a great deal of peatland. The shift from deciduous to coniferous wood for coaling may thus reflect not only a change in forest composition, but also how charcoal production moved from the central parts of the agrarian landscape to distant peatlands and poor soils further out in the outlands. If this interpretation is correct, it can in turn be seen as a sign of the population growth and agrarian expansion of the 18th and 19th centuries, which resulted in a greater pressure on the landscape and increased competition between different activities. Even though northern Scania was still relatively rich in woodlands in comparison to fertile plain areas, deciduous trees in the agrarian landscape also became an increasingly valuable resource here when the population grew. These trees produced acorns, leaf fodder, timber for fences, etc. and may therefore have been protected from coaling.

In conclusion, the history of iron production and coaling and its impact on the landscape in northern Scania can be divided into different phases. Soon after the initial colonization and agricultural expansion in the Early Middle Ages, iron production became part of the local economy. After some centuries it grew in importance and reached a peak during the 16th and early 17th centuries. In addition to ore, the iron production needed a great deal of fuel, and charcoal was produced in large quantities to keep the bloomery furnaces burning. All kinds of deciduous trees were used but there was some preference for birch. During the expansion phase fuel collection and coaling may have gone hand in hand with the agricultural expansion, when woodlands were cleared to give way to pastures, meadows and arable fields. However, in spite of all the clearings and coaling, there was still a great deal of woodland in the area when the period of iron production ended in the 17th century. After the end of the iron production era, charcoal was still produced but the production now changed character. It became a remote activity that took place far way from settlement, using pinewood in particular from peatland vegetation.

MAN-MADE HEATHLAND:
BIRTH AND DECLINE

·········

Heathland in Sweden and abroad

Starting with the early-medieval colonization the uplands of northern Scania were gradually transformed from mainly natural woodland into open or semi-open cultural land, as discussed in previous chapters. The development was not simple and straightforward but rather complex and step-wise, and above all the long-term expansion was interrupted by the Black Death and the crisis and decline that followed during the late 14th and the 15th centuries. After the decline, in spite of the relatively poor climate of the Little Ice Age there was once again expansion with population growth and farm establishment from the 16th century onwards. The agricultural expansion varied in speed and intensity but more or less continued throughout the 17th and 18th centuries and reached a peak and a turning point in the 19th century, at least when it comes to landscape openness. In the late 19th and early 20th century forest plantation and farm abandonment resulted in a reversed landscape development, eventually leading to the much forested landscape of today.

Thus the 19th century, before the start of reforestation, was the century that showed the most open landscape, in northern Scania as well as in many other upland areas of Sweden. It was also the time when the population of the countryside reached its maximum, before the large-scale emigration to cities and overseas to America. The relatively open agricultural landscape of the 19th century can be seen as the end product of a long period of gradually increased agrarian population and increased impact on the landscape and its resources.

An interesting and characteristic feature of the open 19th century landscape was *Calluna* heathland. The investigation area was part of the Southwest-Swedish heathland zone, which stretched from Scania in the south up along the western part of southern Sweden through the provinces of Halland, Småland, and Västergötland to Bohuslän (figure 34a). The vegetation on these heaths was strongly dominated by heather (*Calluna vulgaris*), but also other plants occurred, such as bilberry (*Vaccinium myrtillus*) and lingberry (*V. vitis-idaea*), and grasses such as wavy hair-grass (*Deschampsia flexuosa*) and sheep's-fescue (*Festuca ovina*) (Malmer 1965). Higher vegetation was usually absent, but occasionally there were shrubs of juniper (*Juniperus communis*) and scattered trees of silver birch (*Betula pendula*) and Scots pine (*Pinus sylvestris*). The geographical distribution of heathland coincided fairly well with the parts of southern Sweden that have the highest annual precipitation, i.e. areas that today have more than 800 mm/year (figure 34b). In an international perspective *Calluna* heaths were restricted to the Atlantic zone of Western Europe, from Portugal in the south to the coastland of Norway in the north – a zone characterized by high precipitation, mild winters, and not too hot summers.

Obviously heaths are climate-dependent, but their appearance cannot be explained by climatic factors alone. With the exception of some areas with extreme maritime conditions, such as for instance the Faroes and parts of Scotland (Peglar 1979, Jóhansen 1996), heathland establishment seems always to have been connected to human impact. This is certainly true for Sweden. Here the heaths would not have existed without human-induced deforestation in the first place and they were maintained and kept open by agriculture, or more precisely by grazing, fire clearances, and occasionally by mowing. They were restricted mainly to outlands, in particular on poor, sandy soil, and they were grazed by sheep, goats and cattle. Sheep seem to have thrived on the heaths, while cattle preferred more delicate vegetation such as herbs and grasses when available. To improve grazing the heaths were burnt regularly, usually at 5–10-year intervals (e.g. Malmström 1939). The short-term effect of such heath-fires was that heather was rejuvenated – large and woody plants were burnt down, giving way to new

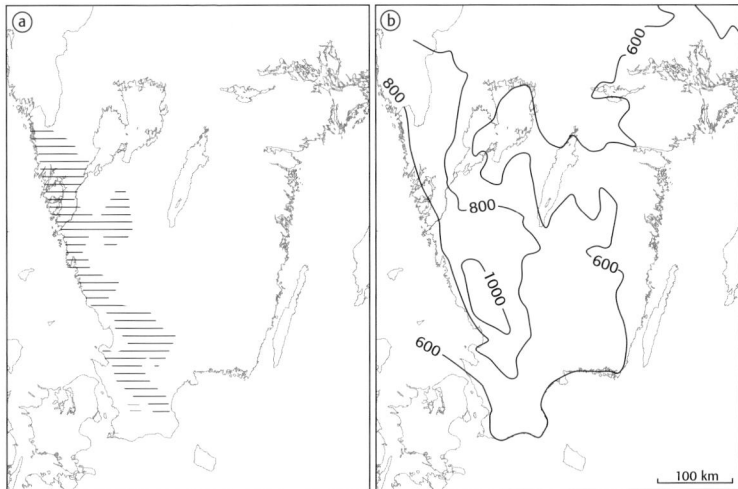

Fig. 34. Maps of southern Sweden showing (a) the distribution of Calluna heathland in 1900 (based on Malmer 1968), and (b) annual precipitation (average for 1961–1990; based on Alexandersson & Andersson 1995).

and more nutritious shoots. The fires also resulted in some years of richer grass and herb vegetation. The long-term effect of the heath-fires, however, was impoverished soils and an even stronger dominance of heather.

During the 18th and 19th centuries, when central authorities started to show more interest in the agrarian economy and in the utilization of natural resources, the heathland management and in particular the heath-fires were condemned and regarded as old-fashioned, inefficient and destructive (Joelsson 2006). This opinion was most strongly put forward during the 19th and early 20th centuries, when tree planting was encouraged (see the chapter *The Return of the Forest* below). But also as early as the 18th century a dislike of heaths and heath-fires was expressed, for instance by Carl von Linné. When he saw burnt *Calluna* heaths during travels in northern Scania in 1749 he wrote that "anyone who understands the laws of vegetation and economy would shiver with terror at the sight of such land-use" [my translation] (Linnaeus 1751, p.89). On the same journey he also passed large heathland in Småland and wrote that "if anyone could find out an easy way to exterminate heather without using fire, he would do the farmers in Småland the biggest favour" [my translation] (Linnaeus 1751, p.54). This negative view has more or less remained in Sweden until the present day, while in other countries, for instance in Norway, *Calluna* heathland is sometimes characterized rather as an ecologically balanced, carefully managed, and efficient land-use system (Haaland 2002). Its main advantage in Norway, Sweden and elsewhere, which is usually put forward, was probably that it offered good winter grazing, in particular for sheep. In this it may have had a major advantage over grassland.

When the heaths had their largest distribution they covered a relatively large proportion of the landscape in western Sweden. In Halland, for instance, which is the province immediately north of the investigation area in northern Scania, it has been estimated that *Calluna* heathland covered 10–30% of the entire province in the mid-19th century (Malmström 1939, Joelsson 2006). Today, almost all the heaths are gone with the exception of a few remnants preserved and managed by nature conservation (Stenström & Forshed 2004). The disappearance of heathland in Sweden is connected to two main processes. One is what may be called the general modernization of agriculture – including for instance land amalgamation and enclosure reforms and the introduction of modern fertilizers – due to which some poor heathland could be turned into arable land. The other and by far most important factor was the introduction of modern forestry, i.e. silviculture (Malmström 1939, Eliasson 2002). During the 19th century the economic value of forest and timber increased and the authorities initiated campaigns to encourage tree plantation, for instance by giving free advice and subsidizing tree seeds and seedlings. The campaigns bore fruit and gradually large areas of heathland and other low-productive agricultural land were turned into forest – in particular pine and spruce forests. By the mid-20th century almost all the heathland in Sweden was gone.

Hence, we have a fairly clear picture of the decline and final disappearance of heathland during the last one or two centuries. When it comes to the early history of heathland, however, the picture is much more unclear. Several Swedish pollen-analytical studies have addressed the question of heathland establishment but they are of shifting quality and together they give a diverse picture. More comprehensive studies on heathland development have been published in Denmark (Odgaard 1994) and Norway (Prøsch-Danielsen & Simonsen 2000), and these may serve as a background for a presentation of the Swedish situation.

On the sandy soils of western Jutland in Denmark and on the coast of Norway there were some small and temporary *Calluna* heaths as early as the Mesolithic. These pre-agriculture heaths were probably connected to fires, either from lightning or induced by Mesolithic hunters. However, it was not until later – after the introduction of agriculture – that more permanent heathland was established due to grazing and fire clearances. In Jutland the earliest permanent heathland formation has been dated at one site to 3000 BC, i.e. Neolithic, but usually somewhat later at other sites. The geographical variation in timing was interpreted as reflecting different soils as well as differences in settlement and land-use intensity. In general, however, the Bronze Age seems to have been the period of most profound heathland expansion in Jutland (Odgaard 1994). In Norway the pattern was rather similar, with a significant expansion of heathland during the Bronze Age. In the coastal zone, heathland establishment and expansion was completed as early as some centuries B.C., while in inland areas it continued throughout the Iron Age until approximately A.D. 1000 (Prøsch-Danielsen & Simonsen 2000). Both in Denmark and Norway it has been concluded that periods of heathland expansion cannot be connected to any specific climatic changes, but instead seem to reflect a gradual and long-term effect of agricultural land-use, in particular grazing and heath-fires.

If we now look at the pollen diagrams from western Sweden, and in particular from the province of Halland and adjacent parts of Scania and Småland (i.e. the central area of former heathland), we get the following picture. Some diagrams show a significant rise in *Calluna* values as early as the Late Neolithic or Bronze Ages. They are from Lake Sämbosjön, Lake Käringsjön, and Lake Iglasjön, which are situated in low-lying areas relatively close to the coast, and from Lake Sandsjön and Exhult, which are situated further inland and at higher elevations. Even though the diagrams show similar increases in *Calluna* pollen, the interpretations of the different authors differ. At Exhult the high *Calluna* values were not interpreted as reflecting grazed heathland, but rather as originating from natural vegetation on the local peat bog (Björkman & Ekström 2003). Also at Lake Käringsjön doubts about the interpretation were put forward as the first *Calluna* expansion coincided with a local expansion of bog peat vegetation (Björkman & Persson 2005). However, at Lake Sandsjön the early *Calluna* rise was interpreted as a reflection of grazed heathland (Thelaus 1989), in spite of the fact that the site is situated in the most bog-rich part of Sweden, and that the sampled lake borders a peat bog. What may support the interpretation of grazed heathland at Lake Sandsjön is that some other grazing indicators, such as *Rumex* (sorrel, etc.), increase at the same level in the diagram, but on the other hand the same level also shows a strong increase in *Sphagnum* (bog moss) spores, which indicates an expansion of natural bog vegetation.

This discussion reflects a general problem concerning the interpretation and dating of heathland establishment based on pollen diagrams. The problem is that heather (*Calluna vulgaris*) thrives in two very different environments: one is grazed heathland on dry ground, one is natural peat bogs. Heather does not thrive on minerotrophic fens, but when fens gradually develop into nutrient-poor ombrotrophic bogs their surface is invaded by heather, which soon becomes a dominant component of the local vegetation. Because most peatland sooner or later develops from fen to bog, pollen diagrams may show increasing *Calluna* values even without any heathland at all.

Of the diagrams mentioned above that show an early (i.e.

Bronze Age) increase in *Calluna*, Lake Sämbosjön provides the strongest indication of early grazed heathland, as it is situated in an area with not many bogs (Digerfeldt 1982). However, the *Calluna* frequencies from that lake are not very high to begin with, and it is not until further up in the diagram, at layers from the Middle Ages and later, where really high values are reached. Finally, Lake Iglasjön, the fifth site with early *Calluna* expansion, shows a similar development as Sämbosjön, with a first increase during the Bronze Age and higher values later, during the last few centuries (Wallin 2004).

In contrast to the diagrams with early *Calluna* expansion, there are several diagrams from the same area where the first *Calluna* rise has been dated to considerably later periods. Such diagrams are, for instance, the ones from Kullaberg (Björkman 2001) and Lärkesholm (Ljung 2003) in northwestern Scania, Köphult (Björkman 2003) in southwestern Småland, and Baggabygget (Björkman 2005), Trälhultet (Björkman 2000a), Yttra Berg (Sköld 2006), Bocksten (Björkman 1997a), and Lake Gyltigesjön (Wallin 2004) in Halland. All these sites do not show a first significant *Calluna* expansion until the Middle Ages or later, and in most of them the expansion waits until the last two or three centuries. In addition there are also a number of sites where the pollen diagrams show low *Calluna* values all the way up to the top.

In conclusion, although it cannot be entirely excluded that some early records of *Calluna* heathland may in fact be misinterpretations of natural bog vegetation, there seems to have been heathland establishment at some sites in southwestern Sweden as early as the Bronze Age. However, this early heathland expansion was not as vigorous and large-scale as in Jutland and Norway, but rather restricted and patchy in character. In many areas there were open pastures at the time but they were still dominated by herbs and grasses. The main heathland expansion in southwestern Sweden seems to have started considerably later, in connection with the medieval agricultural expansion, and to have been strongest during the last two or three centuries. A later and weaker heathland expansion in

Sweden than in Jutland and coastal Norway may be due to somewhat less maritime climatic conditions.

Local heathland development and underlying causes

The new pollen diagrams from the investigation area in northern Scania presented in this book support the general picture of a late heathland expansion. The four sites show some variation but together they give a clear picture of the regional heathland development in this upland area (figure 35).

Starting with the southwestern site Östra Ringarp, we find a relatively weak but continuous *Calluna*-signal. Even before the first significant deforestation and local farm establishment in the early Middle Ages, *Calluna* shows a continuous curve reaching 1–2% of the pollen sum. The frequency does not increase after the farm establishment, in spite of a significant increase in Poaceae (grass) and other grazing indicators, but remains at 1–2%. It suggests that the *Calluna* pollen at these levels in the Östra Ringarp diagram does not reflect grazing but rather peatland vegetation in the area. Slightly higher and more even values are noted from the 16th century onwards, but it is not until approximately the early 19th century that *Calluna* shows a stronger increase. At the same level there is also a significant increase in Poaceae and other grazing indicators, as well as in cereals and weeds of arable land. Therefore, the *Calluna* rise in the 1800s can with certainty be interpreted as heathland establishment due to increased human impact. It should be noted, however, that the *Calluna* values are not very high in comparison to those of Poaceae, so in addition to heathland there were still pastures with richer grass and herb vegetation.

In the uppermost part of the diagram from Östra Ringarp, *Calluna* declines abruptly and at the same level there is an increase in the percentages of *Pinus* (pine) and *Picea* (spruce). It reflects reforestation and the introduction of silviculture. It is worth noting that Poaceae does not decline in the same way as *Calluna*, but instead continues with relatively high values to

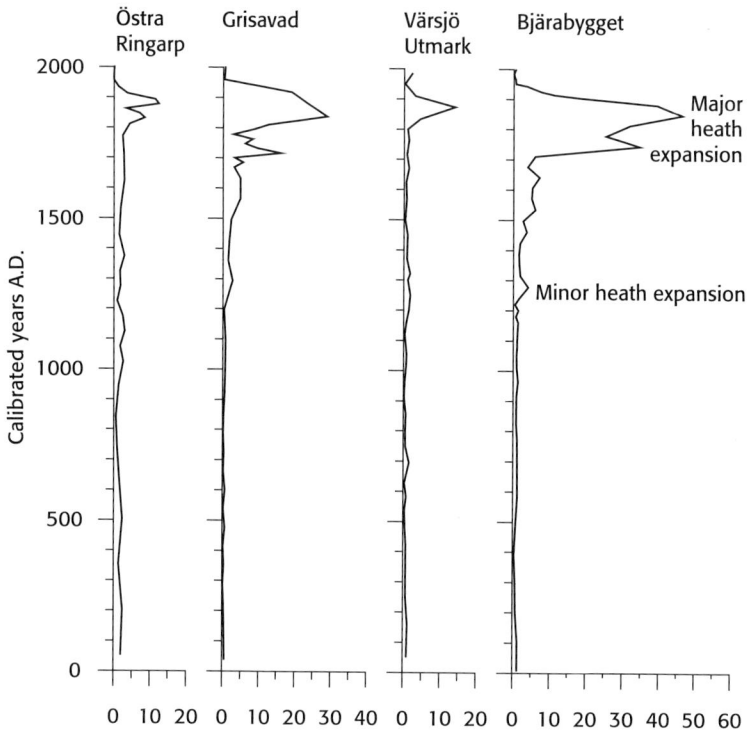

Fig. 35. Calluna vulgaris (heather) pollen graphs for each site. The graphs show percentages of total pollen.

the top of the diagram. Obviously, poor heathland was chosen for forestation, while some grassland was kept open and remained in use as pastures for the farm that still exists at the site today.

The next site, Grisavad, is similar to Östra Ringarp in that the early-medieval agricultural expansion does not result in any increase in *Calluna* pollen, in spite of a strong increase in Poaceae (grass). The first increase in *Calluna* is found somewhat further up, at a level dated to the 13th century, but the values remain relatively low. A stronger increase in *Calluna* started later, during the 16th century, i.e. in connection with the farm establishment and agricultural re-expansion after the late-medieval decline. However, it was not until the 18th and in particular during the 19th century that heathland became widespread. At these levels in the diagram *Calluna* reaches really high values (20–30%). The same levels show high values also of cereals and weeds, but a gradual decreasing trend of Poaceae (grass). We can conclude that heathland achieved maximum distribution during the 19th century in the Grisavad area and that the expansion was partly at the expense of grassland.

The diagram shows an abrupt decline of *Calluna* in the uppermost part, where *Pinus* (pine) and *Picea* (spruce) percentages increase. Also here heathland, but not only heathland, was forested in the early 20th century. The local crofts were abandoned and all cultural land connected to them was turned into forest.

At the third site, Värsjö Utmark, a slight increase in *Calluna* is noted at the level of farm establishment, dated to the 12th century. The increase is weak, however, in particular when compared to the strong increase in Poaceae (grass) at the same level. No heathland was established during the Middle Ages, in spite of strong indications of grazing, but some scattered plants or patches of heather may have occurred in the pastures. A more significant increase in *Calluna* is dated to the 19th century, where it reaches 10–15% of the pollen sum. These high values reflect a relatively short-lasting phase of heathland, connected with agricultural expansion and to the establishment of some crofts in the area. What is interesting to note is that the expansion starts with a very distinctive rise in the Poaceae curve, and that in the next phase Poaceae declines while *Calluna* increases. It seems as if the initial clearing – reflected in the sharp decrease in *Betula* (birch) – resulted in the expansion of grassland that very soon developed into heathland. Possible explanations to this change will be discussed below. The heathland at Värsjö Utmark lasted for approximately a hundred years or less, before it was transferred to pine and spruce forest.

Bjärabygget, the north-easternmost site, is similar to Värsjö Utmark in that the diagram shows a weak increase in *Calluna* percentages at the level of farm establishment. At Bjärabygget this level is dated to the 13th century. Also here the *Calluna* values during the Middle Ages are much lower than those of Poaceae, and they are too low to be interpreted as heathland. The values increase further in the 16th century and increase very sharply in the 18th century, to reach maximum values of 40% in the 19th century. Before the *Calluna* expansion in the 18th century, pastures were characterized by grassland with a minor component of heather, but during the 18th century vast heathland was established at the expense of grassland, and also at the expense of birch woodlands, which seem to have been cleared at the time. High percentages of *Pinus* pollen may indicate that there were scattered pines in the heathland.

When the grazing and management of heaths at Bjärabygget ended in the late 19th or the early 20th century, they were overgrown by birch in the first phase and then by spruce. The latter was probably planted while the birch expansion reflects a natural succession.

Taken together, the four pollen diagrams presented here give a rather unanimous testimony of the regional heathland development. To begin with, there was no prehistoric heathland. This result came as no surprise, as our investigations have shown that there were not any prehistoric agrarian settlement or land-use in the area, except for herding and extensive wood pasturage. What perhaps may be more surprising is that there was no or very little heathland during the Middle Ages either, in spite of well-established agriculture and permanent grazing. Clearings and grazing created and maintained pastures during the Middle Ages, but they were characterized by grasses and herbs rather than heather (*Calluna*). In connection with the 16th-century expansion, increasing abundance of heather is noted at some sites, but still not enough to indicate real heathland. The only period from which the diagrams provide clear evidence of *Calluna* heathland is during the 18th and 19th centuries. At that time heathland seems to have been rather widespread and more common than grassland at all the investigated sites. Finally, all the diagrams in their upper parts catch the final transition from heathland to coniferous forest, i.e. the end of the heathland epoch and the introduction of modern silviculture, which links the diagrams to the present situation (see also figure 43).

Hence, the heathland development in the investigation area was rather late. An interesting question is: why did heathland not develop earlier? Grazing had been going on in the same areas since the early Middle Ages, but it was not until during the 18th and the early 19th centuries that heathland developed. Something must have changed, it seems, for instance in the land-use practices or in the natural pre-conditions.

One possible explanation may be a more frequent use of fire. In three of the four diagrams high percentages of microscopic charcoal match high *Calluna* percentages, and hence show stronger indications for the use of fire during the 18th and 19th centuries than during earlier periods. However, if fire was used only to rejuvenate heather and to increase the grass and herb component of the heathland vegetation, it would not have been used in the first place if there had not already been a problem with heather. From this line of reasoning the establishment and development of heathland should not only be explained as an effect of heath-fires, even though such fires may have reinforced and speeded up the process once it had started. However, fire may also have been used for other purposes, for instance in slash-and-burn cultivation or to clear shrubs and brushwood in pastures, which to some degree may have facilitated the expansion of heather.

Another important factor behind the heathland expansion, in addition to fire and grazing, is soil leaching. As heather is a tolerant and competitive species on nutrient-poor soil, the expansion of heathland was favoured by soil degradation. Heather vegetation itself creates an acid moor layer, which leads to acid soil and increased leaching of nutrients, which in turn makes heather more competitive. This positive feedback, in which heather improves the conditions for itself, is one important factor behind the heathland expansion. But similar to fire it does not seem to explain the initial heathland establishment, and it certainly does not explain why no heathland developed during the Middle Ages.

To understand the timing of heathland establishment in northern Scania we also have to look at two other factors: population rise and climate change. Even though estimations of population size before the 17th century are tentative, there is no doubt that the population of Sweden, Denmark and many other countries was greater during the 18th and 19th centuries than ever before (e.g. Livi Bacci 2000, Andersson Palm 2001). As mentioned above, population in the countryside of northern Scania and other upland areas peaked in the 19th century, and a larger population meant increased agricultural production and greater pressure on natural resources. When it comes to pastures, increased livestock may have resulted in higher grazing pressure, leading to nutrient depletion and impoverished soil, and, as a consequence, heathland establishment.

Thus the poor character of the late grazing grounds in comparison to their medieval predecessors may be explained by soil exhaustion caused by increased agricultural impact and population growth. But climate may also have had something to do with it. Medieval agriculture was aided by the favourable climate of the Medieval Warm Period, but later, in particular during 16th and 17th centuries, the climate deteriorated significantly (figure 30). This cold period, known as the Little Ice Age, may have affected the soils in a negative way. Even though we do not know the exact character of the climatic conditions of the Little Ice Age in a local perspective, it is plausible that low temperatures and increased humidity speeded up the natural process of soil leaching. When human impact and grazing pressure increased in the 18th and 19th centuries, soils were more vulnerable and susceptible to depletion than before the Little Ice Age.

To sum up, heathland in the investigation area and in many other parts of southwestern Sweden expanded rather late, mainly during the 18th and 19th centuries, in spite of a much longer land-use history. The heaths were the product of a combination of grazing, heath-fires, poor soils and maritime climatic conditions. The timing of heathland establishment can tentatively be explained by strong population growth in combination with soil exhaustion. The latter process may have been driven by high grazing pressure but was probably also reinforced by a cool and wet climate in the Little Ice Age.

19TH CENTURY CROFTS

. . . .

Croft expansion in Sweden

During the late 18th century and the early 19th century a large number of smallholdings were established in northern Scania and in many other parts of the Swedish countryside. In Sweden they are called *torp* (e.g. Elgeskog 1945). The best English translation is probably crofts; a term which was originally used for smallholdings in Scotland but may also be used in a more general way. Terms such as these, however, are difficult to translate, not so much because of a lack of appropriate terms, but because the systems and the realities that they refer to have been different in different countries. (And to add to the confusion the Swedish word *torp* has a wide range of definitions and its meaning has varied over time.) In this book the word croft will be used for any smallholding that during the 18th and 19th centuries was established in an outland area, and which was not an independent tax-paying unit but belonged to the farm on whose outland it was established, and to which it had to pay a rent in money, products, or labour obligations.

A typical croft in the uplands of southern Sweden was a small farm with a wooden house, one hectare of arable and meadow, two cows, and some pigs and sheep. It was situated in the outland, sometimes far away in the forest. The crofter supplemented his small-scale agriculture with other work, such as forestry, carpentry, peat cutting, etc., and he also did day-labour on the farm to which the croft belonged. It was also common that the crofter was a soldier in the regular army.

Crofts were an important component of the agricultural expansion during the late 18th and 19th centuries. Similar to heathland (which was discussed in the previous chapter) the crofts can be connected with the strong population rise during that period (figure 36). The Swedish population tripled over less

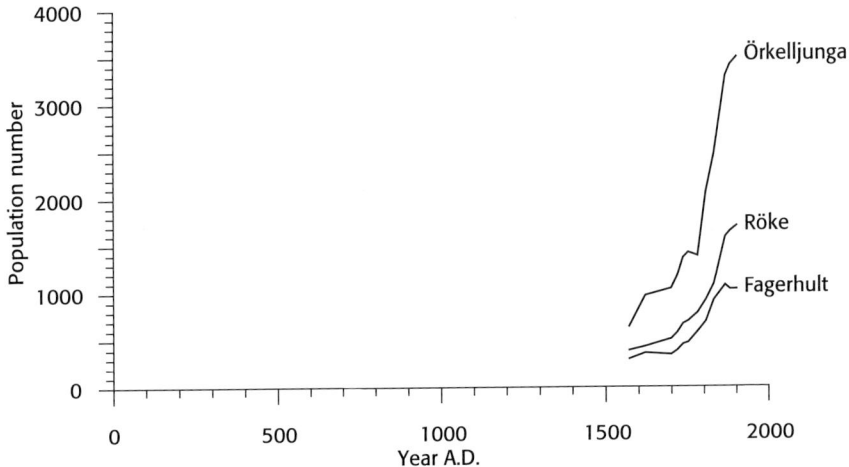

Fig. 36. *Population numbers 1571–1900 in the three parishes of the investigation area in northern Scania (data from Andersson Palm 2000).*

than two centuries, from 1.4 million in 1700 to 4.2 million in 1870 (Gadd 2000). But in a reciprocal way the croft expansion should not be seen only as a reflection of population rise, but also as a prerequisite of it. Thanks to the crofts even distant parts of the outland could be cleared and put under the plough. Agriculture associated with the individual crofts may have been small-scale – the arable was usually only one tenth of that of a freehold – but due to the large number of crofts, the process of croft establishment contributed significantly to an overall increase in agricultural production.

But the croft movement should not only be associated with agrarian expansion. It also meant that outland areas were populated, and as a consequence different outland activities expanded and increased in importance. As crofters had to supplement their agriculture with other work to make a living, they were an important labour resource that came to good use in forestry, peat cutting, coaling, etc. Eventually, during the late 19th and the early 20th centuries, the crofters also found jobs within industry. The crofts may therefore be described in two different ways: both as an expansion of the traditional agricultural economy and as a link and a transition to the modern, industrial economy (Nordström 1996).

Many crofts had a rather short history. A few developed and expanded to full-scale farms, some survived as houses for workers, and some were turned into summer houses. The vast majority, however, were abandoned in a large-scale abandonment process that lasted during the late 19th and the early 20th century. Today the faint remains of crofts can sometimes be found, in particular in forested areas. Ruins of the simple houses, together with some stone walls and scattered clearance cairns – resting in the shadow of dense coniferous forest – are touching reminders not only of hard work of past generations, but also in some areas of a densely populated countryside that is hard to imagine today.

Local crofts in a long-term perspective

Crofts are an important part of our recent past, and many Swedes are only two generations away from their crofter ancestors. Much has therefore been written about crofts and the life of a crofter, in particular by historians and genealogists, but during the last few decades there has also been an increasing interest among archaeologists. Some croft remains have been excavated, and in spite of the existence of written records that are sometimes detailed, it has been shown that archaeology may provide new information and contribute with new aspects (e.g. Welinder 1992, Rosén 1999, Lind et al. 2001). Within the present project in northern Scania, croft remains at the pollen-analytical sites Grisavad and Värsjö Utmark were subject to excavations. Also at Bjärabygget croft remains were found close to the pollen-analytical sampling site, but outside the excavation area.

Pollen analysis, in addition to archaeology, may give an important contribution to the study of crofts by providing information on the land-use and on the surrounding landscape. Its most important advantage, however, is that it has the possibility of putting the crofts in a long-term context of landscape and settlement history. Focus in the following presentation will be on the results of pollen analyses, but with some supplementary information from archaeology and written sources.

At Grisavad, on a sandy plateau covered by dense spruce forest and bordering a peatland, there were the visible remains of a croft, which were subject to a small-scale excavation within the project (A. Knarrström 2003, in prep.). One of the aims of the pollen diagram from Grisavad was to study the land-use associated with this croft, and to put the visible and excavated remains into a long-term landscape perspective. The peat core used for the analysis was therefore taken close to the croft and its arable, only 20 metres from the edge of the peatland (figure 37).

On the site at the time of investigation, there were the remains of a small dwelling house, outhouses, a well, and some stone walls. The buildings had probably been timber-framed with a thatched or turfed roof, but all that remained were the stones of wall foundations and a collapsed chimney together with some cellar pits. In close connection to the house remains were also field terraces, clearance cairns and some ancient pathways. Based on the distribution of stone walls, terraces, and stone-cleared ground, it could be estimated that the croft had approximately 0.5 hectares of arable fields. All the visible remains on the site were mapped and a few square metres were excavated. Archaeological finds were pottery, china, glass, nails, a comb, locks, coins, buttons, bones, etc. Most of these finds could be dated to the 19th century, but some of them could perhaps also be from the 18th or 17th centuries. Additional information about the croft was collected from written sources, old maps, and from personal communication with Bo Hellström, who is the grandson of one of its last inhabitants.

The croft was situated on the outland to the hamlet Östra Ringarp, to which it belonged. In the same area there were also a number of other crofts, at least during the 19th century. When the crofts were established is, however, not easy to tell. The place-name Grisavad seems to be associated not only with the

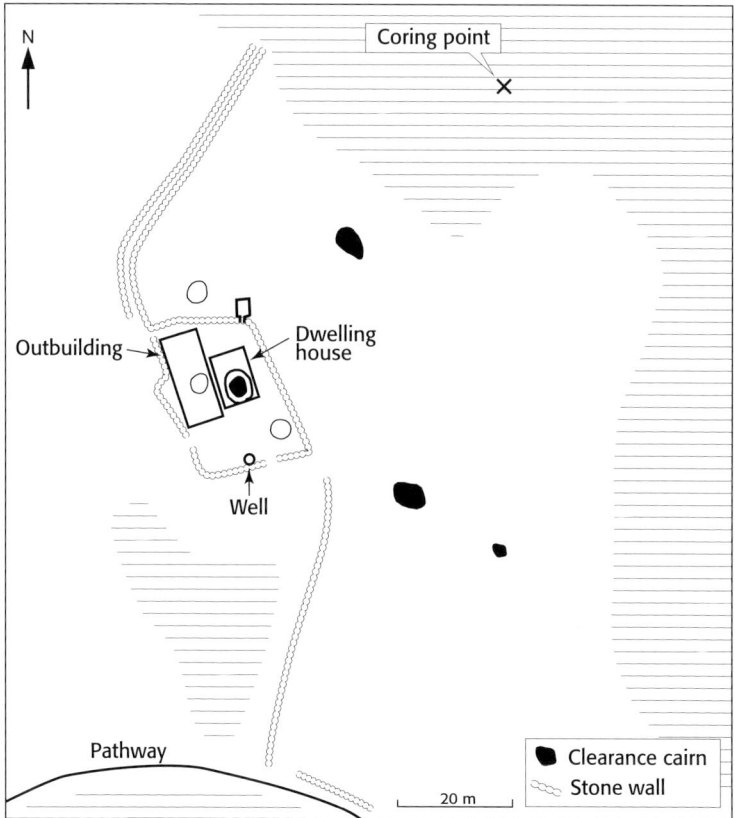

Fig. 37. Map of the Grisavad site with the remains of an abandoned croft. Hatching indicates peatlands. The coring point for pollen analysis is indicated.

investigated croft, but also to this part of the outland. Settlement at Grisavad is mentioned in written sources from the 19th century, and also from 18th and 17th centuries, with the earliest mention in the Swedish Civil Survey from 1671 (Sw: *jordrevningsprotokoll*) (Skansjö in prep.). It is not possible to say if any of these early mentions refer to the investigated croft, but obviously some settlement was established in the area as early as the late 17th century.

In contrast to the fragmentary information about the establishment, we have a clearer picture of the last phase and the final abandonment. From parish registers (Sw: *kyrkoböcker*) several people can be connected to the croft during the 19th century, of which the last ones were the soldier Magnus Brink and his family. They left in 1901, which was therefore the date of abandonment. When the croft was revisited by an earlier inhabitant in 1911 it was in ruins. The spruce forest that covers the place today was planted in the 1930s.

If we now turn to the pollen analysis, the upper part of the diagram shows strong indications of agriculture – both cultivation and grazing – during the 17th, 18th and 19th centuries (figure 38). The diagram has a local character and the agriculture that it reflects can be ascribed to the Grisavad smallholdings in general and to the investigated and closely situated croft in particular. It gives evidence of the cultivation of rye (*Secale*), barley (*Hordeum*), wheat (*Triticum*), oat (*Avena*), buckwheat (*Fagopyrum esculentum*), and flax (*Linum usitatissimum*), which perhaps is a greater diversity of crops than one would expect from a poor croft settlement. To this we may add potato, which is not possible to prove by pollen analysis, but is known to have been cultivated on the site (one of the small fields at the croft was called the Potato Field). What is interesting to note is that the cultivation indicators in the diagram show as high frequencies during the late 17th and 18th centuries as they do throughout the 19th century. If the croft and its arable were established in the 19th century we would expect increasing frequencies. We may therefore tentatively conclude that the croft was established as early as the late 17th century. Hence, it is still possible that the mention of Grisavad in 1671 actually refers to the investigated croft.

In contrast to the cultivation indicators, the frequency and character of grazing indicators change throughout the period. In the early phase, i.e. during the 17th and 18th centuries, grassland dominated, but during the 19th century there was a strong expansion of *Calluna* heathland. The heathland expansion probably reflects increased grazing pressure connected to population

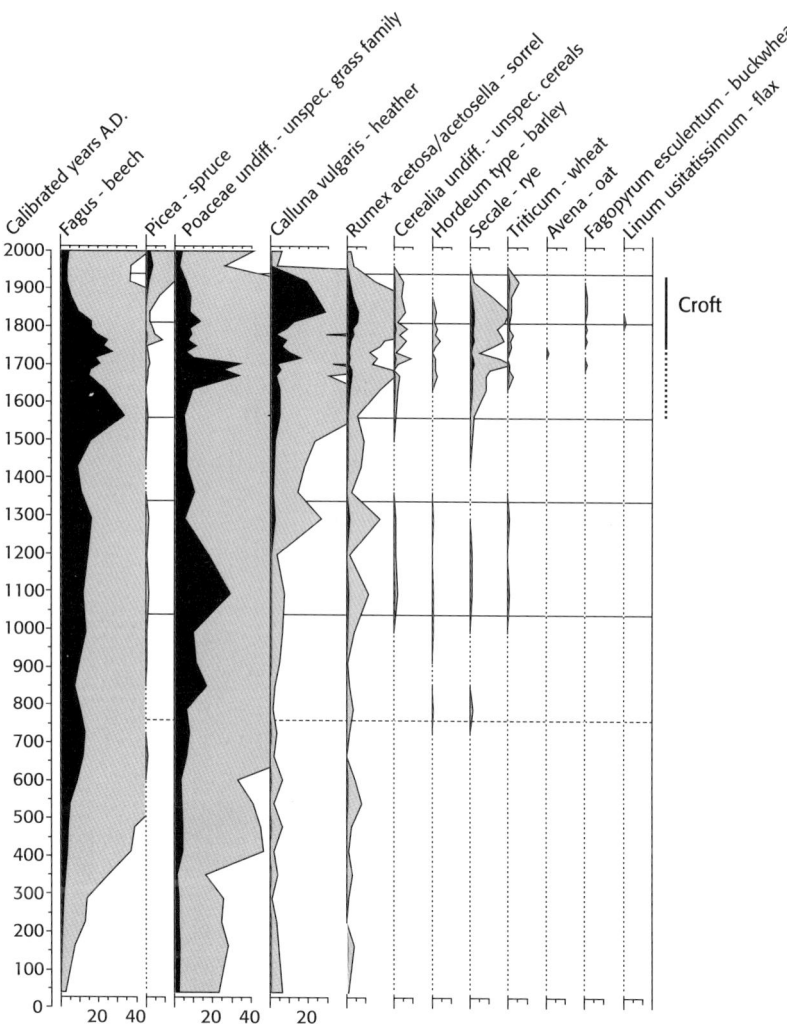

Fig. 38. Pollen diagram from Grisavad with selected taxa. The section that reflects the land-use of the nearby croft is indicated. (For details see caption to fig. 9.)

145

rise and possibly to further croft establishment in the area (see discussion on heathland in the previous chapter). A decrease in *Fagus* (beech) pollen shows that the expansion of heathland was partly at the expense of beech forest.

In the uppermost part of the diagram, the sharp decrease in cultivation indicators and *Calluna* reflects the abandonment of the croft, while the increase in *Picea* (spruce) and *Pinus* (pine) reflects the forestation that followed. These recent events are not possible do date in detail by pollen analysis or radiocarbon dating, but they seem to fit in well with the dates of the croft abandonment (1901) and the forest planting (1930s) that were obtained from other sources.

Thus the diagram from Grisavad gives information about local vegetation and land-use and also contributes to the discussion of croft establishment and abandonment. Its most important contribution, however, is that it provides evidence also of an earlier phase – of a farm that was established somewhere in the vicinity as early as the Early Middle Ages but was abandoned in the 14th century. (This result was presented and discussed in previous chapters.) Beech forest that had expanded on the abandoned land during the late-medieval decline was cleared when agriculture was re-introduced in the late 16th and 17th centuries. Land that was once used for cultivation and grazing was taken into use again. The early-medieval phase was not known from the historical record and it was not captured by the small-scale excavation – it only speaks to us through the pollen diagram.

From Grisavad we now move to Värsjö Utmark – the other pollen-analysed croft-site within the project (figure 39). At Värsjö Utmark there were also the forest-covered remains of a croft, with house foundations, stone walls, clearance cairns, etc., which were mapped and partly excavated (A. Knarrström 2004, Wennström in prep.). The ruins of the dwelling house were clearly visible but situated outside the exploitation area and could therefore not be excavated (all excavations within the project were restricted to the exploitation area for a new motorway). In addition to the dwelling house there was an outhouse with a waste layer. The layer was partly excavated and yielded

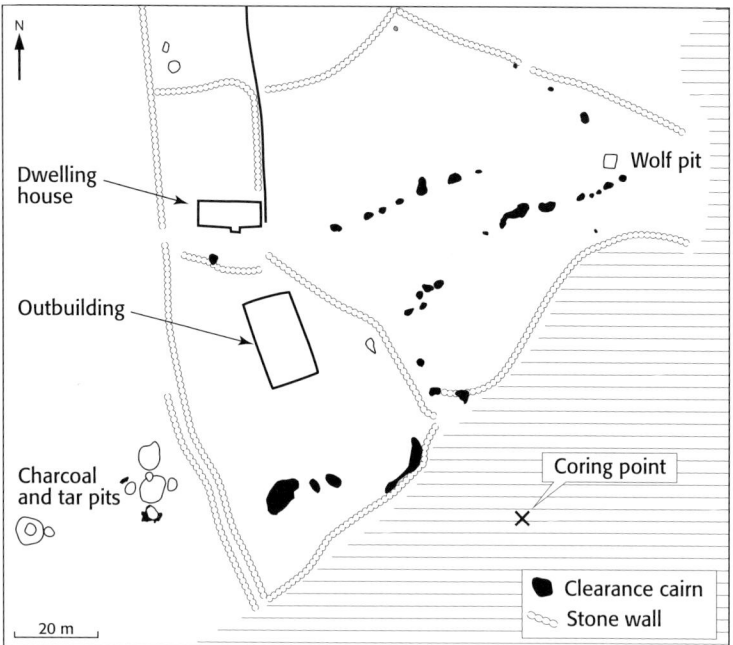

Fig. 39. Map of the Värsjö Utmark site with the remains of an abandoned croft. Hatching indicates peatlands. The coring point for pollen analysis is indicated.

many finds – pottery, glass, buttons, iron nails, etc. – from the 19th century and possibly earlier. Charcoal from clearance cairns and a group of kilns and tar pits were radiocarbon-dated, and the dating gave interesting results. Some of them could be referred to the croft, as expected, while three of them (one clearance cairn, one tar pit, and one kiln) gave medieval dates. Obviously the croft at Värsjö Utmark, similar to the one at Grisavad, had a medieval predecessor, and here the medieval phase not only appeared in the pollen diagram but also in radiocarbon dates from the excavation.

The written records and other documents from Värsjö Utmark are fragmentary but add some information. The croft was situated in the outland of Värsjö, and was indicated on a map of Värsjö from 1834. In documents supplementary to the map the croft is referred to as *Rosts täppa* (*Rost* is a personal name and *Rosts täppa* simply means Rost's arable plot). The person Rost who is mentioned is Troed Rost, a soldier who was born in Värsjö in the late 18th century, and together with his wife Nilla inhabited the croft during the 19th century. Probably the croft was finally abandoned when Nilla died in the 1870s. We do not know if there had been inhabitants before the Rost family.

With the aim of reflecting the local land-use of the investigated croft, a core for pollen analysis was taken in the adjacent peatland, only 20 m from the edge and approximately 30 m from the croft's arable. In spite of poor time-resolution due to secondary compaction of the upper part of the peat sequence, it is possible to identify the croft phase in the pollen diagram (figure 40). At a level tentatively dated to the early 19th century there is a significant increase in cereal pollen, in particular of *Secale* (rye). At the same level there is a sharp decrease in *Betula* (birch) and an increase in Poaceae (grass) and some grassland herbs and weed, such as *Rumex acetosa/acetosella* (sorrel), *Potentilla* type (tormentil, cinquefoil), *Galium* type (bedstraw), and *Artemisia* (mugwort, wormwood). These changes in the pollen diagram are interpreted as the establishment of the nearby croft, i.e. the very croft that on the map from the early 19th century was referred to as Rost's place. Before the establishment of the croft the area was covered by light birch woodland with some remaining *Juniperus* (juniper) shrubs, and the sharp decrease in *Betula* pollen reflects how this birch woodland was cleared when the croft was established. As discussed in the previous chapter some of the grassland in connection to the croft soon developed into *Calluna* heathland.

Thus, the upper part of pollen diagram reflects land-use of the nearby croft during the 19th century. However, the cereal record shows that some cultivation was going on somewhere in the vicinity even before the local birch woodland was cleared, i.e. from the 16th century onwards. The cereal record from this period is weak but continuous, and it probably reflects arable

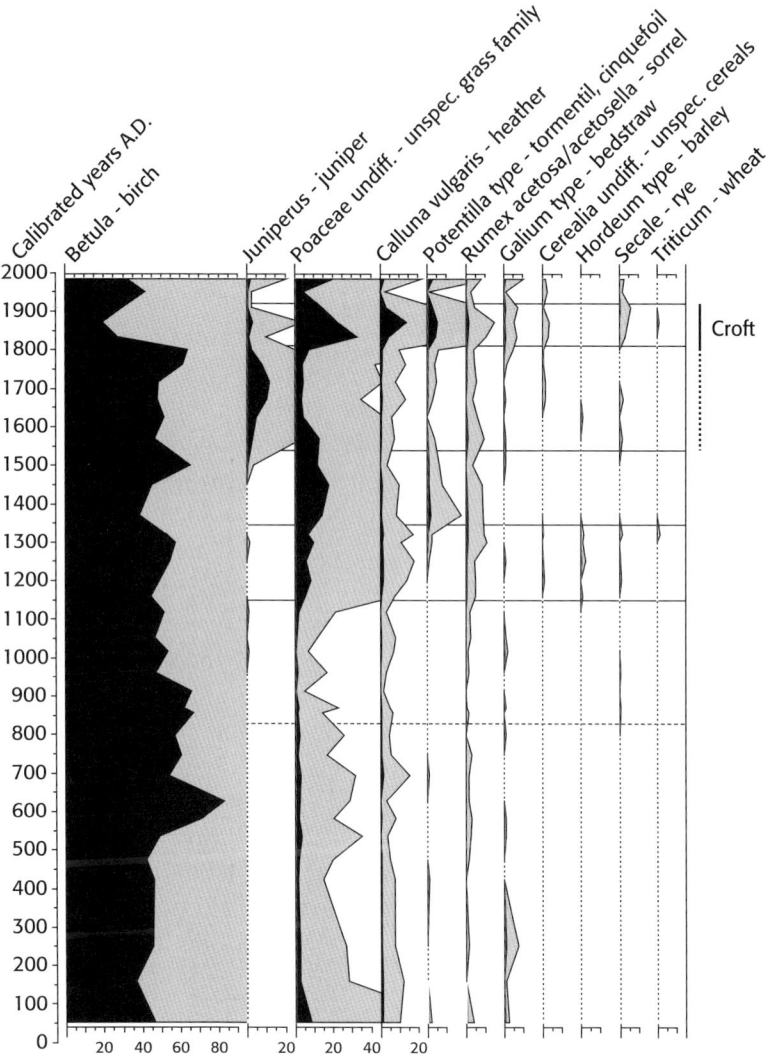

Fig. 40. Pollen diagram from Värsjö Utmark with selected taxa. The section that reflects the land-use of the nearby croft is indicated. (For details see caption to fig. 9.)

and smallholdings at some distance from the sampling point but still in this part of the Värsjö outland.

But there had also been a much earlier settlement phase. Radiocarbon dates from the excavation indicated early-medieval activity on the site, as mentioned above, and this result was confirmed by the pollen analysis. The diagram gives evidence of a local farm, which was established somewhere in the vicinity during the 12th century and was abandoned during the 14th century, as discussed in previous sections. So, rather like Grisavad, crofts and other smallholdings in the outland of Värsjö do not represent the first agricultural settlement in the area, but rather a re-establishment after the late-medieval decline.

In the uppermost part of the diagram from Värsjö Utmark, finally, we see the croft abandonment and introduction of silviculture – pollen frequencies of cereals and grazing indicators decrease while *Pinus* (pine), *Betula* (birch), and *Picea* (spruce) increase. In contrast to Grisavad the area around the Värsjö Utmark site was not completely forested. All the settlement and most of the agricultural land were abandoned, but some parts are still kept open by grazing, and three hundred metres away there is a field that is still cultivated.

To sum up, the pollen analyses at Grisavad and Värsjö Utmark have succeeded in reflecting land-use connected to the investigated crofts. In general the results are in agreement with the results of the archaeological excavations and with the information we got from written sources and old maps. The analyses have also contributed to the discussion on timing of croft establishment, even though the pollen diagrams as such can never say if the land-use was managed by crofters or freeholders. The diagrams also reflect the late croft abandonment and reforestation, which gives a link to the present landscape. Their most important contribution, however, is by showing that the crofts at these two sites were preceded by medieval settlement and agriculture, which had been abandoned in the 14th century. Obviously, abandoned agricultural land was re-occupied by the later crofts, even though the buildings and other constructions may not necessarily have been on the same sites.

Croft establishment and abandonment

Historical sources provide rather detailed and accessible information about crofts and their inhabitants, and the information gained from pollen analysis should be seen as only complementary. In particular the absolute chronology of the pollen diagrams, which is based on radiocarbon dating, is too coarse to provide any detailed information on the timing of, for instance, the abandonment process. The pollen diagrams are still important, however, as they facilitate comparison between the croft phase and earlier periods, such as the Middle Ages, from which historical data are scarce.

The pollen diagrams from Grisavad and Värsjö Utmark indicated that cultivation was practised in those areas from the late 16th century or the early 17th century onwards (the medieval phase is not discussed here), but also that new crofts were established in the early 19th century and that grazing pressure gradually increased. It fits in well with the general picture of the croft expansion in southern Sweden, according to which it started as early as the 17th century but was intensified later, in particular during the early 19th century (Nordström 1996). The pollen diagrams can only detect vegetation and land-use – and indirectly also settlement – but they cannot tell which settlement is to be referred to as a croft and which is not. However, the settlement and arable that were inferred from the diagrams at Grisavad and Värsjö Utmark were situated in the outland, relatively far away from the respective hamlets and their infields. Therefore it is justified to assume that they were crofts or some similar type of non-freeholders. The conclusion may therefore be that the first smallholdings, probably crofts, were established in the outland of Östra Ringarp and Värsjö (the investigation sites Grisavad and Värsjö Utmark, respectively) as early as the late 16th century or early 17th century, and that additional crofts were established during the late 18th and early 19th centuries. This process was a reflection of a general agricultural expansion in this region and also in Sweden as a whole.

In comparison to the phase of establishment and expansion, the process of croft abandonment is less complicated and as it is not so distant in time it is also better documented. From a great many sources we know that most crofts in Sweden were abandoned during the late 19th and the early 20th centuries, and that general picture is supported by the investigations and analyses presented here. The abandonment was usually abrupt, but it could also be step-wise as some crofts continued to be inhabited by workers when the agriculture that was connected with them had been closed down.

Regardless of their juridical status, most crofts were to be regarded as small farms, with arable, hay meadows and pastures. From the perspective of long-term land-use history they marked the peak and eventually also the end of the traditional agricultural society, which expanded from the 16th century until the 19th century. The succeeding abandonment of them – in northern Scania and elsewhere – was a reflection of the large-scale depopulation of the Swedish countryside, during which more than one million people emigrated from Sweden to America and even more moved to nearby cities and industries (Morell 2001). It was certainly one of the most profound abandonment phases ever witnessed in Sweden and the reforestation that followed was probably the most dramatic landscape change ever. Croft ruins in dense coniferous forest are evidence of this dramatic phase in Swedish history, and they are important reminders of an ever-changing landscape.

THE RETURN OF THE FOREST

·········

Pine and spruce expansion

In connection with the abandonment of poor agricultural land
and the general depopulation of the countryside during the
late 19th and early 20th centuries there was strong forest expansion. In particular coniferous forest expanded, with dominating tree species being spruce (Norway spruce *Picea abies*;
here simply called spruce) and pine (Scots pine *Pinus sylvestris*; here called pine), and still today dense coniferous forest
very much characterizes the landscape. Some deciduous forest
also expanded, e.g. with beech (*Fagus sylvatica*), but to a
much less extent. The strong dominance of coniferous trees
distinguishes the late reforestation period from the earlier one
that followed upon the Black Death in the 14th century, during
which in particular deciduous trees such as birch and beech
expanded.

The late reforestation and expansion of coniferous forest is
clearly reflected in the new pollen diagrams from northern
Scania as well as in several earlier published diagrams from
other parts of southern Sweden (e.g. Björkman 1996, Lindbladh 1998, Lagerås 2002b). The graphs in figure 41 show
how the pollen frequency of both *Pinus* (pine) and *Picea*
(spruce) increases significantly in the uppermost part. According to the absolute chronology of the diagrams the main reforestation can be dated approximately to the late 19th and the
early 20th century. A more precise date cannot be obtained
based on pollen analysis, partly due to the wide calibration intervals of the last three hundred years, and partly due to interpolation difficulties in the uppermost few centimetres of the
peat sequences. However, the approximate date is in good
agreement with the general picture of the reforestation process
that we have from written sources (Malmström 1939).

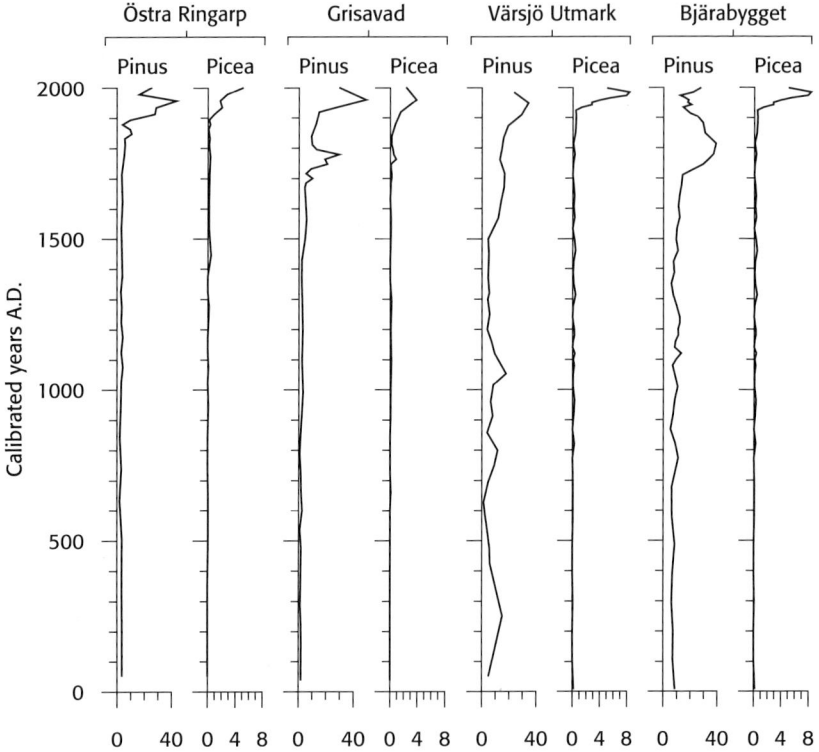

Fig. 41. Pinus (pine) and Picea (spruce) pollen graphs for each site. The graphs show percentages of total pollen. Note that different scales are used on the x-axes for Pinus and Picea.

If we now compare the graphs of pine and spruce in figure 41, we find that both species expanded in the late 19th and early 20th centuries, but we also find an interesting difference. Spruce pollen is very much restricted to the uppermost levels, while pine pollen in some diagrams also shows high frequencies at levels from earlier periods. The difference reflects different history and migration patterns. Spruce, in contrast to pine and most other tree species, expanded from north to south through middle and southern Sweden, and it was a very late immigrant to northern Scania. Even if some scattered trees and small stands may have existed early on, the general spruce expansion showed a front-like pattern, and according to compilations of several pollen diagrams the "spruce front" reached northern Scania only one or two centuries ago (Björkman 1996, Giesecke & Bennett 2004). It has been concluded that the southwestern limit of the present day distribution of natural spruce forest runs through the investigation area in northern Scania (Lindquist 1959), even though it is almost impossible to define its natural distribution today when most spruce forest is planted and climate is rapidly changing.

Several authors have discussed possible factors behind the spruce expansion and to what degree its present distribution is controlled by climate (e.g. Webb 1986, Sykes & Prentice 1996, Bradshaw et al. 2000). Most seem to agree that its distribution to the south and to the west is limited by warm climate and in particular by too mild winters, and that its expansion during the last few millennia on a broad scale can be seen as a response to a general trend of climate cooling. However, today we experience global warming due to the greenhouse effect and we may therefore expect spruce forest to be pushed back north again, and such a future scenario has in fact been suggested based on computer simulations (Bradshaw et al. 2000). It has also been argued that the severe effect of recent storms indicates that planted spruce forest in southern Sweden is not in pace with the rapidly changing climate (Bradshaw et al. 2000). Even though it is hard to imagine today, when spruce is so dominant, the spruce phase may in the long run turn out to be only a short episode in the landscape history of northern Scania. However, regardless of uncertain future scenarios, there is strong evidence that spruce forest arrived very late in northern Scania and that picture is supported by the new diagrams presented here.

In comparison to spruce, pine has a much longer history in southern Sweden. It was one of the first tree species to immigrate after the deglaciation and it has been here now for about ten thousand years. However, similar to spruce, it shows increasing pollen frequencies in the uppermost part of the diagrams and obviously it has been more common during the last few centuries than ever before. In the investigation area in northern Scania it was present before the late expansion, which is clearly shown in figure 41 by the graphs from Värsjö Utmark and Bjärabygget. Sometimes low frequencies of pine pollen are explained by long distance pollen transport, but in the investigation area there is no doubt that pine really was present locally (evidence is provided for instance by numerous finds of macroscopic pine charcoal; see *Charcoal data* in Appendix 2). Pine has always been present in the area, but it has mainly been restricted to peat bogs and to poor sandy soil, which is also its main habitat today.

In some of the graphs in figure 41, for instance the one from Värsjö Utmark, *Pinus* shows a gradual increase starting as early as the 16th century. While expansion during the 19th and 20th centuries certainly was due to human-induced forestation on dry ground, expansion during the 16th–18th centuries may tentatively be explained by peatland development. In general peat bogs in the area have expanded during the last one thousand years and as a consequence suitable habitat for pine has also expanded (the peatland development in the area has been subject to a separate study; Lagerås 2003a, Edlund 2004). Possibly the pine expansion that started in the 16th century reflects enhanced bog development in the course of the Little Ice Age.

To sum this section, pine and spruce have different histories, but both of them showed strong expansion in northern Scania during the late 19th and the 20th centuries. Their late expansion may be seen as the birth of the present-day landscape, which is very much dominated by coniferous forest.

The process of reforestation and the introduction of silviculture

On a broad scale the distribution of pine (*Pinus sylvestris*) and spruce (*Picea abies*) is dependent on climate, but of course it is also dependent on land-use. Spruce in particular is a dominant and shade-tolerant tree, which may establish itself and expand at the expense of other tree species in a natural forest (Björk-man 1996). However, in northern Scania this was rarely the case. Most of the pine and spruce forest that was established in the investigation area did not replace other forest but rather replaced open agricultural land. The expansion of these tree species was part of a significant forestation process in which open land was turned into forest. This is evident from the pollen diagrams, in which the late increase in pine and spruce pollen is accompanied by a sharp decrease in most open-land taxa.

If we examine the pollen diagrams in more detail we also find indications of what type of open land was turned into forest. What is most striking is the entire disappearance of *Calluna* heathland. In all the diagrams the frequency of *Calluna* pollen decreases sharply at the same level as *Pinus* and *Picea* increase. Poaceae undiff. (grass) also declines in some of the diagrams, but not as strongly as *Calluna*, and in some diagrams it does not decline at all. This relationship indicates that poor heathland in particular was forested while some grassland was kept open by continuing agricultural land-use, i.e. grazing and mowing.

Not only heathland was turned into forest. In some places there was total forestation of the entire agricultural landscape. This was the case for instance at the investigated croft site Grisavad, where all pastures, meadows, and arable, together with the habitation area, were forested. Grisavad may be seen as a local example of a common trend during the late 19th and early 20th centuries, according to which forestation was made possible by the abandonment of crofts and farms and by the fact that large areas of poor land were taken out of agricultural production.

But what was the actual process behind forest expansion? In some cases it may simply have been natural overgrowing of abandoned land, i.e. forest establishment due to spontaneous tree-seed dispersal and germination when grazing, heath-fires or other agricultural management ceased. In other cases it may have been deliberate sowing or planting. The latter process was probably by far the most common one. A study from the province of Halland showed that some natural forestation by pine on heathland started back in the early 19th century, but that deliberate forestation by sowing or planting increased in importance and became most common during the late 19th and throughout the 20th century (Joelsson 2006). The situation in northern Scania was probably similar. When comparing the tree species, we may also expect that there was a difference between pine and spruce (figure 41). Pine was already established in the area and scattered pine trees grew on heathland and on other open land, in particular on poor soil. If grazing pressure decreased or management ceased, such pine trees would have resulted in natural forestation by seed dispersal. Spruce, on the other hand, was not present in the area, or at least it was rare, and was therefore dependent on sowing and planting.

As mentioned above the forest that was established was primarily a coniferous one. The only deciduous tree taxon that also expanded significantly was birch (*Betula*). Birch is a light-demanding early-succession tree, and it probably colonized recently abandoned land by natural dispersal. The diagram from Östra Ringarp shows an interesting relationship between the pollen graphs of *Calluna*, *Betula*, *Pinus*, *Picea*, and *Fagus*, which may shed light on the forestation process at least in that area (figure 42). Immediately after the decrease in *Calluna* and some other open-land indicators there is an increase in *Betula*, *Pinus*, and *Picea* pollen. However, a more strong increase in *Picea*, together with an increase in *Fagus*, starts at the next level where *Betula* and *Pinus* decrease. Hence, the forestation may be separated into two phases – a first one characterized by birch, pine, and spruce and a second one characterized by spruce and beech. The first phase may tentatively be interpreted as a combination of natural succession and seed dispersal on the one hand and sowing and and planting on the other, while the later phase may be interpreted as the result of sowing and planting only. Note also that hazel (*Corylus*), which was abundant in the area before the heathland expansion, did not re-expand, possibly as an effect of soil degradation during the heathland phase.

The transition from agriculture to silviculture in Sweden was a process that was encouraged and supported by central

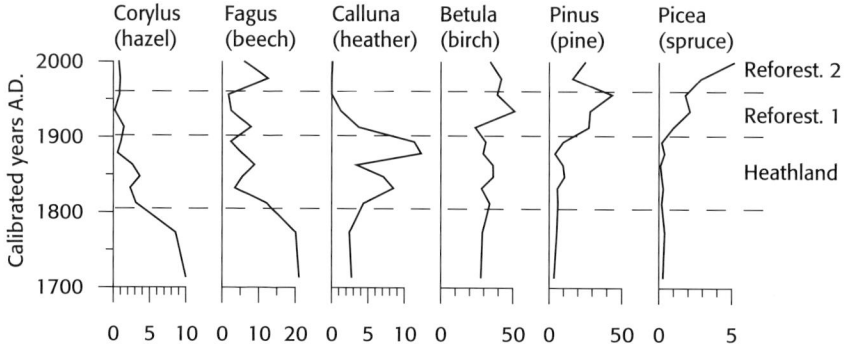

Fig. 42. Pollen diagram from Östra Ringarp with selected taxa. The reforestation after the heathland phase is divided into two phases. Note that different scales are used on the x-axes.

authorities (Malmström 1939, Eliasson 2002). In particular the successful achievements of forest science in Germany were regarded as an inspiring model, and from the early 19th century onwards authorities stimulated forestation and the introduction of modern silviculture in Sweden. Landowners received guidance and information and they were also given financial support and subsidies. Thus, the forestation process should not be seen only as a natural response to the depopulation of the countryside and to agricultural abandonment, but also as a result of political decisions and campaigns.

From the perspective of silviculture and forestry industry there is no doubt that it was a successful campaign. But in hindsight we may also conclude that many environmental values were lost on the way. Heathland, which was a characteristic feature of western Sweden, disappeared completely, and so did large areas of species-rich grassland and hay meadows. In marginal areas in particular a complex and diverse agricultural landscape was replaced by monotonous tree plantations. Most modern coniferous forests in southern Sweden today are monocultures with low biodiversity. In some places small remnants of past vegetation types have survived, even in dense

Fig. 43. Three maps of the investigation area in northern Scania showing the distribution of deciduous forest (light shading) and coniferous forest (dark shading) at three different times: (a) 1684, (b) 1865, and (c) today. The maps are redrawn and presented here in the same scale to facilitate comparison, even though the original maps were of different scales and with different degree of accuracy. Map (a) was based on Gerhard Burmann's map of Scania (approx. 1:250,000), map (b) was based on Generalstabskartan (1:100,000), and map (c) was based on Fastighetskartan (1:10,000). The pollen investigation sites of the present study are indicated in (c). Based on the redrawn maps, the distribution of open land, deciduous forest, and coniferous forest was calculated with a GIS computer programme:

	1684	1865	2000
Open land (%)	55	77	27
Deciduous forest (%)	37	18	11
Coniferous forest (%)	8	5	62

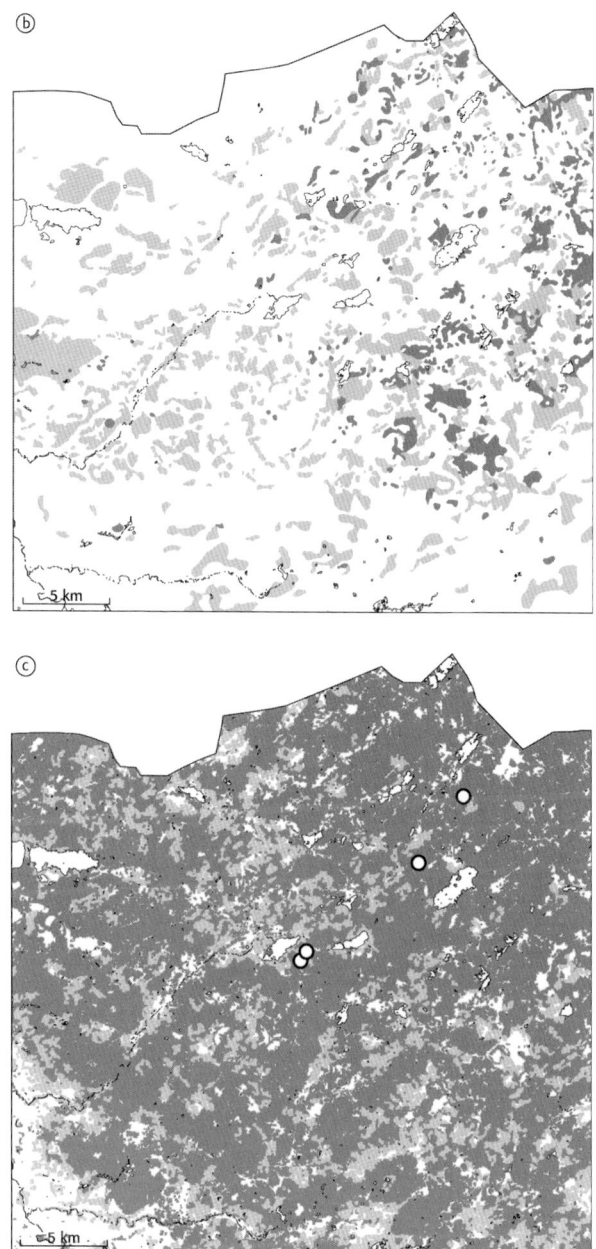

spruce forest. Shrubs of, for instance, lilac (*Syringa vulgaris*) or hawthorn (*Crataegus* sp.) may be found by croft ruins and other settlement remains, but they are of course just tiny fragments of what has been lost in the course of forestation.

The immense landscape change due to forestation during the late 19th and the early 20th centuries is evident in a series of maps presented in figure 43. The maps show the distribution of forest cover in the investigation area in northern Scania at three different times, i.e. in the years 1684, 1865 and today (see figure caption for details). According to the map from 1684, there was a forest cover at that time of approximately 45%, of which the majority was deciduous forest. In 1865 – when the landscape was probably more open than ever before or after – forest cover was only 23%, still with a strong dominance of deciduous trees. Today there is a forest cover of 73% and the major part (85%) of it is coniferous. Between 1865 and today total forest cover has more than tripled and coniferous forest cover has increased by as much as twelve times (from 5 to 62%). (Note that the area used for calculation in figure 43 includes a part of the agricultural lowland in the bottom-left corner; if it had been restricted to pure upland, the degree of forest cover would have been even higher.)

The calculations of past forest cover are based on what can be identified as forest in the different maps. Deciduous forest is separated from coniferous, but the maps do not say anything about forest structure. We know that the present coniferous forests have the character of dense plantations aimed at maximizing timber production. But during the time periods reflected in the earlier maps, forest structure was certainly very different from today. Early woodlands were used not only to produce timber, but also for wood pasturage, pollarding, fuel collection, coaling, tar production, slash-and-burn cultivation, etc., and consequently they must have had a more open and diverse structure than modern forest plantations. In contrast to the sharp borders between forest and open land today, we may also expect early landscapes to have been greatly characterized by gradual transitions from open land to forest – for instance because of gradually decreasing grazing pressure at greater distances from in-fields and settlement. In conclusion, the landscape change between 1865 and today must have been even greater than what is evident from the calculated change in forest cover presented above.

In a broader perspective the large-scale reforestation witnessed in northern Scania was typical for marginal uplands in southern Sweden. Some plain areas were also forested, in particular the ones that had poor sandy soil, but most fertile plains remained open or even increased their openness. Uplands may always have had more forest than fertile plains, at least since the Bronze Age (Lagerås 1996a), but this regional difference was now strongly increased, leading to a more polarized landscape than before. Thus, in connection with the reforestation process of the late 19th and the 20th centuries, we may identify increased polarization at two different levels, both on a local level, where the difference between forest and open land has become greater and the borders sharper, and on a regional or supra-regional level, where agriculture is greatly concentrated to fertile plain districts while marginal upland areas are used mainly for silviculture.

SYNTHESIS

· · · · · · · · ·

Regional development and temporal diversity

Previous chapters in this book have dealt with different landscape periods and processes during the last millennium. The interpretations and discussions have been based on new palaeoecological studies in northern Scania, in combination with results from archaeology, human geography, and history, and comparisons have been made with earlier studies from other parts of southern Sweden. The chapters have been more or less in chronological order, starting with the early-medieval colonization and ending with the recent forestation that created the present landscape. Depending on the subject and on the data at hand, the different chapters have a somewhat different character. Sometimes the discussion has been rather broad, with a supra-regional or North-European perspective, and sometimes more detailed with a strong focus on the local landscape and on land-use changes at farm level. This deliberate jumping between different scales may have been confusing, but it reflects my ambition to put local results in a larger perspective and to highlight the historical connections between the local society and the outside world. It also reflects my belief, that, to be meaningful, interpretations of any past vegetation and land-use have to be made in the light of what we know about society of the time.

The empirical basis for the book has been four local pollen diagrams. Using local pollen diagrams from the same region in this way, instead of one regional pollen diagram, has several advantages. First of all, the fact that local diagrams are more site-specific than regional ones makes it easier to compare the results with other site-specific data, for instance from archaeological excavations or from the written record. It also enables conclusions to be drawn about individual farms. As farms are the key-units of the agricultural landscape, interpretations at farm level make a good basis for discussions of land-use practices and landscape change. Based on the site-specific investigations of the present study, it was for instance possible to show how later crofts were established on the abandoned land of early-medieval farms. Another advantage of using several local studies is of course the possibility of studying local diversity within the region. Several examples of such local diversity, in both space and time, were discussed in previous chapters. For instance it was possible to show that the early-medieval colonization was a step-wise process moving from the southwest to the northeast.

Thus, by comparing the four local pollen diagrams, it has been possible to discuss details and to highlight local diversity and complexity. But together the diagrams also give a strong regional picture of the landscape and settlement development (figure 44a). This regional development has also been discussed in previous chapters, but for clarity its main characteristics will be summarized here (for references and detailed discussions see previous chapters):

The investigated upland area in northern Scania was used only for extensive wood pasturage and herding during most of prehistoric time. There are no known prehistoric graves in the area and the pollen diagrams are strongly dominated by tree pollen but with some grazing indicators. From the pollen data and a series of radiocarbon-dated hearths it seems as if continuous herding and wood pasturage started during the Bronze Age, but may have started earlier.

The first signs of cultivation are a few cereal pollen grains from the Viking Period. They reflect small-scale, temporary cultivation and may be seen as a prelude to the medieval colonization. If these clearings were meant only to be temporary, or if they reflect an agricultural expansion and settlement establishment that for some reason was never completed, it is not possible to say. However, the landscape remained forested throughout the Viking Period and it seems as if no permanent settlement was established. No grave-fields from the Viking Period are known in the area.

It was not until the Early Middle Ages that a more profound agricultural expansion reached this upland area. Starting in the southwest and gradually reaching further up into the more elevated parts in the northeast, permanent farms with arable fields and pastures were established in a colonization process, which lasted from the 11th to the 13th century. Possibly farm establishment continued all the way up to the mid-14th century, when expansion turned into decline, however the last farm establishment documented by the pollen diagrams was dated to the 13th century. It is difficult to quantify land-use and landscape openness from pollen data, but simply the fact that local farms were established at all the four sites during the Early Middle Ages, indicates that the expansion was vigorous and that early-medieval farm settlement was widespread. Cultivated crops were rye, barley, and wheat, and at one of the sites also hemp and flax. Pastures were characterized by grasses and herbs, and they did not develop into heathland in spite of several hundred years of grazing. The local early-medieval society was to a large degree an agricultural society, in particular in its initial phase, but iron production and possibly other outland use grew in importance later on. (The colonization process will be further discussed in a separate section below.)

The agricultural expansion that characterized the Early Middle Ages came to an end in the 14th century. At two of the four investigated sites cultivation stopped, which probably reflects abandonment of the local farms. At one of them, abandoned agricultural land was to some degree reforested, while at the other it was kept open by grazing. Land-use at the other two sites (those that did not show signs of abandonment) lingered on as before but without expansion. The result from the investigation area is more or less repeated in a compilation of a larger number of pollen diagrams from southern Sweden (figure 44b). Similar to the results from northern Scania the compilation shows that the 14th and 15th centuries was generally a period of stagnation and regression. I interpret this period of agricultural decline and abandonment as a direct or indirect reflection of the Black Death, which hit these areas in 1350.

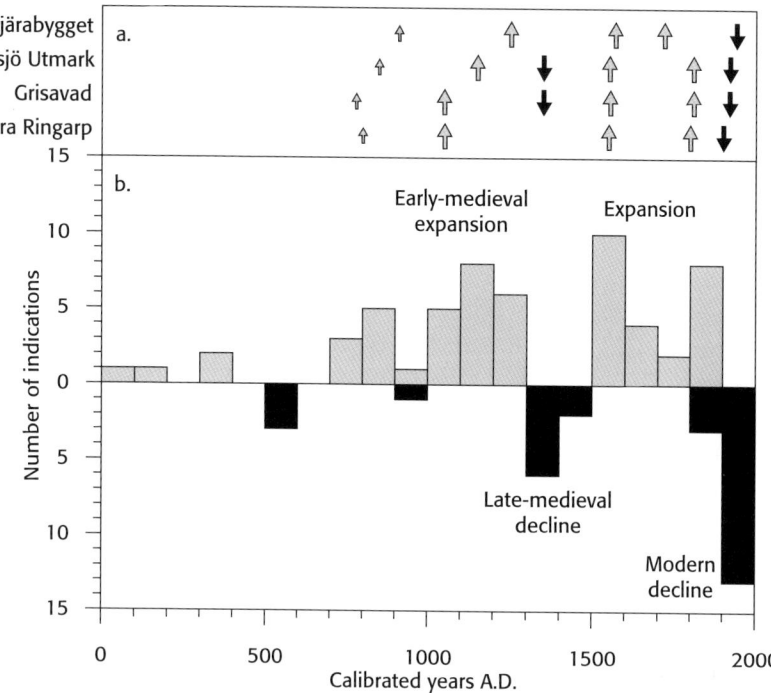

Fig. 44. *Indications of agricultural expansion (grey) and regression (black) during the last 2000 years in (a) the four pollen diagrams from the investigation area in northern Scania, and in (b) 20 pollen diagrams from upland areas in southern Sweden. Based on fig. 24 and 26.*

(The abandonment will also be further discussed in a separate section below.)

A new period of agricultural expansion started in the 16th century. Cultivation and settlement were re-established on abandoned sites, and farms that survived the crisis now expanded. The expansion was strong and it was connected to a significant population rise. As many farms had survived the crisis, the expanding post-medieval society was obviously rooted in its medieval predecessor, but to some degree it was probably also restructured. Iron production now became large-scale and a more complex local economy was developed.

Expansion continued and the entire period from the 16th to the 19th century may be characterized as a long expansion period. However, after the 16th century it seems as if expansion slowed down and the next phase of strong expansion did not occur until the late 18th and the early 19th centuries. Similar to the 16th century expansion, the agricultural expansion of the early 19th century was connected to a strong population rise. The pollen diagrams show expansion of arable of which some can be connected to the many crofts that were now established in outland areas. The same crops as during the Middle Ages were still grown, but now with the addition of buckwheat and oat. Pastures were characterized by grassland to begin with, but gradually they developed into poor heathland. This change was probably the combined effect of high grazing pressure and soil deterioration in the course of the Little Ice Age. However, heathland expansion was rather late, and it was not until during the late 18th and the 19th centuries that heathland was widespread.

After a population peak in the mid-19th century a large-scale depopulation of the countryside started. More than one million Swedes emigrated to America and even more moved to nearby cities and industries. The abandonment process was accompanied by the introduction of modern silviculture, and very large areas of agricultural land were forested. This landscape change is clearly recorded in the four pollen diagrams. While earlier woodlands were dominated by beech, oak, birch, and other deciduous trees, the modern ones show a strong dominance of spruce and pine. Forestation and depopulation were particularly thorough in marginal areas such as in the uplands of northern Scania. In such areas it was the most dramatic landscape change that had ever occurred.

To sum up, the cultural landscape development of the last millennium in the uplands of northern Scania was characterized by two major expansion periods and two periods of regression. The first expansion period started with some temporary clearings in the 9th century but was more profound during colonization and farm establishment in the 11th–13th centuries. The second expansion period started in the 16th century and

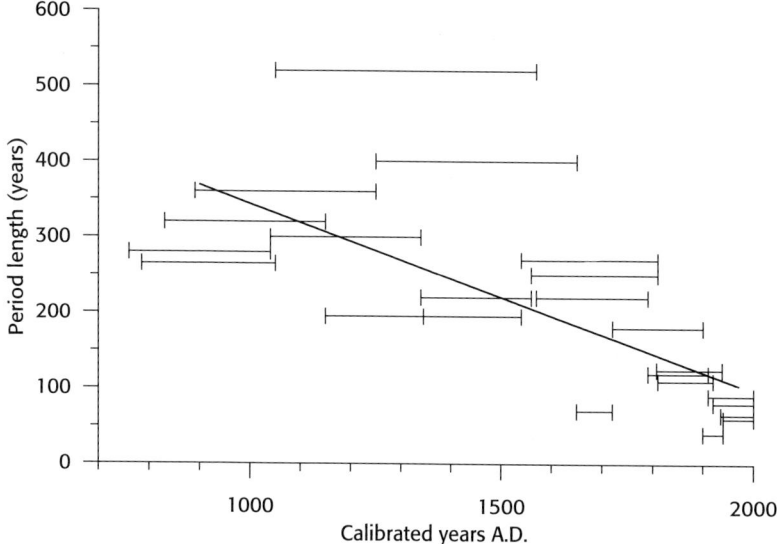

Fig. 45. Diagram showing the relationship between period duration and age during the last 1300 years. The periods presented
are the pollen zones of the four pollen diagrams of the present
study (cf. Appendix 2). Each horizontal bar represents a pollen
zone, which age and duration can be read on the x-axis and y-
axis, respectively. The long straight line shows a linear regression
of the relationship between duration and age, based on the mid-
points of each period ($y = -0.25 * x + 590$, $R^2 = 0.52$).

continued, although with shifting intensity, until the 19th century. Thus, both expansion periods were gradual and long-
term, lasting for several centuries. In contrast, the regression
periods were shorter and more abrupt. In particular the process of large-scale abandonment and reforestation of the late
19th and early 20th centuries was very abrupt, but the earlier regression period – the late-medieval decline – was also relatively sudden. The resolution of the pollen diagrams does not allow any finer pinpointing in time than to century level, but if
the suggested connection to the Black Death is correct, most
abandonment probably happened during the late 14th and the
early 15th centuries.

In addition to the picture of an interplay between expansions and regressions, the cultural landscape development may also be characterized as a series of landscape periods of gradually shorter duration. In general, landscape periods were longer – i.e. significant changes occurred less frequently – during prehistoric times than during later periods. It is evident from the fact that the early-medieval expansion in the investigation area was preceded by a very long period of extensive wood pasturage and herding, which had lasted for more than two thousand years. However, within the time interval of the last one thousand years there is also an obvious trend of gradually shorter landscape periods, and hence more frequent changes, by time (figure 45). It reflects a general trend of accelerating landscape change, probably due to the combined effect of technological development and population rise.

The colonization process

The new results on the early-medieval colonization of northern Scania are particularly interesting, not least as much of this process pre-dates the earliest written records. When the written sources became more detailed in the late 13th century, for instance with the provincial laws, the early-medieval expansion had been going on for some centuries. Even though more detailed historical records exist for some areas (Myrdal 1999a), it is generally difficult to get a picture of the early-medieval farm establishment and settlement pattern based on written evidence. Also from the archaeological record, it has turned out to be problematic to identify rural settlement from the Early Middle Ages, in particular in forested uplands. The situation has improved during the last few decades, partly due to the increased scientific attention that has been paid to clearance cairns and other agricultural remains, but still the information is very fragmentary. Clearance cairns are sometimes dated to the Middle Ages but very few of the rural buildings belonging to them have been subject to thorough excavation and dating (however, see for instance Bjuggner & Rosengren 1999, Lindman 2004, Hansson et al. 2005). The state of knowledge may to some degree reflect the archaeological situation in Sweden, where much focus is on prehistoric time (or on medieval towns), but simply also the fact that rural medieval settlement is hard to identify in the field.

The difficulty in finding rural settlement from the Early Middle Ages is evident also from the present project. Apart from the castle and the Romanesque church in Örkelljunga, both dated to A.D. c. 1300, there is little physical evidence of medieval settlement in the area. Clearance cairns have been dated to the Early Middle Ages, and so have iron production sites and tar pits, but no remains of buildings have been excavated. However, in a similar environment south of Markaryd, some kilometres to the north of the investigation area, archaeologists from the Småland Museum recently excavated the remains of a farm (Åstrand 2006). Pottery and radiocarbon dates showed that it was established in the late 13th century and abandoned in connection with the late-medieval decline during the 14th or early 15th century. The only visible remains prior to excavation were clearance cairns and slag heaps, and the actual building remains were not discovered until the soil was stripped and the stony ground was carefully cleaned. The difficulty in identifying the house illustrates why so few medieval farms have been found and documented in the forested uplands of southern Sweden.

From this background, the new pollen diagrams from northern Scania provide an important contribution to our understanding of the early-medieval expansion. They show the character and timing of the expansion and the landscape change, and, indirectly, they reflect medieval settlement. In particular permanent cultivation, inferred from continuous cereal records, has throughout the book been used as an indication of permanent farms, and was used to establish a chronology for the medieval expansion.

Marginal areas in general are sensitive to changes in society and they are therefore suitable for studies of the medieval expansion. This is true also for the investigation area in northern

Scania. However, the investigation area turned out to be particularly suitable because of the absence of earlier settlement. Most other parts of the forested uplands of southern Sweden show clear evidence of Iron Age settlement, but in spite of careful surveying and numerous excavations no traces of such premedieval settlement were found in the area. Hence, we are not only dealing with medieval expansion, but also with colonization.

As discussed in a previous section, the rim of the upland, i.e. areas to the southwest of the investigation area, was colonized as early as the Roman Iron Age, approximately A.D. 200. The next step was taken considerably later, during the Early Middle Ages, when agriculture and settlement expanded towards the northeast, into upland areas that had never been settled before (with the exception of very early Mesolithic sites). The timing of the early-medieval colonization is reflected in the pollen diagrams, which indicate farm establishment at Östra Ringarp and Grisavad during the 11th century, at Värsjö Utmark during the 12th century, and at Bjärabygget during the 13th century. In addition, the farm south of Markaryd, which was mentioned above, was established in the 13th century. Probably, both the Roman expansion and medieval expansion originated in the agricultural plains to the southwest, but the medieval one may also to some degree have originated in the coastal plains to the northwest, i.e. in the province of Halland (figure 46).

Thus, the results suggest a smooth and gradual colonization process in which settlement and agricultural land expanded step by step across the investigation area in a northeastward direction. In the book *Colonization of unfamiliar landscapes*, two major patterns of colonization are identified, namely streaming and advancing front patterns (Rockman 2003, see also Anthony 1990). Streaming is when colonizers move long distances from known areas to new ones, leaving the areas in between uncolonized. It may be used to characterize, for instance, the American gold rush and other similar situations in which natural resources, vacant jobs, etc. in specific areas have

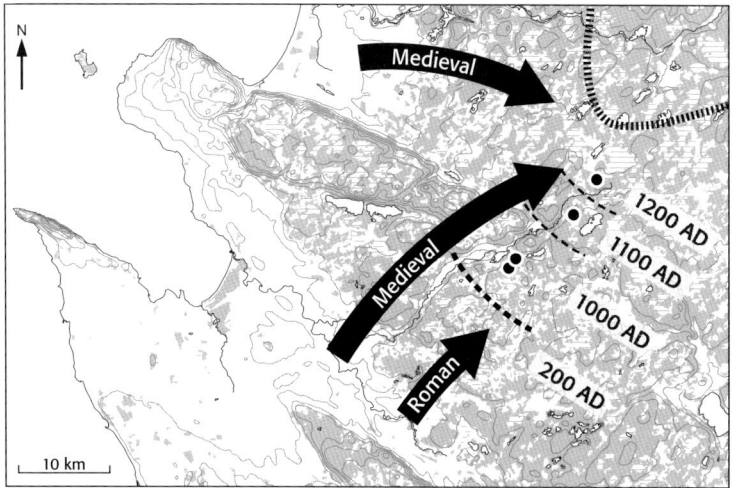

Fig. 46. Map showing the Roman Iron Age expansion and the more profound medieval expansion and colonization in the uplands of northern Scania. Based on fig. 19.

functioned as pull factors. The advancing front pattern (or wave-of-advance) represents regular and short-distance movements into new areas directly adjacent to the previously known ones. It has been used to describe, for instance, the initial peopling of Australia and the Neolithic expansion across Europe.

Although the picture is fragmentary, the new data presented here suggest that the early-medieval colonization of northern Scania had the character of an advancing front rather than streaming. New farms were not established far away in the unknown, but, as it seems, in relatively close connection with already existing ones. It gives the impression of a gradually expanding society, rather than brave pioneers on isolated farms in the wilderness. The advancing front pattern also fits in well with the interpretation that the initial expansion was mainly agricultural, and that iron production and other outland use became important parts of the economy somewhat later. The earliest iron production in the area was dated to the late 12th century but it was not until during the late 15th or 16th centuries that it became large-scale. If the bog ores were an important pull factor behind the initial colonization, we would perhaps expect a streaming pattern, in which colonizers moved more directly to the large peatlands in the northeast. Although there probably would have been some demand for timber and other upland resources, it is in my opinion, more likely that the colonization was primarily driven by population rise, shortage of land, and other push factors in the areas from which the colonizers originated.

Mechanisms of abandonment

From expansion we will now turn to the opposite – periods of decline. According to the results and interpretations presented in this book, there have been two major periods of decline (i.e. periods of agricultural regression and abandonment) during the last one thousand years. One is the large-scale agricultural abandonment and reforestation during the late 19th and early 20th centuries, which is well known from many sources, and the other one is the late-medieval decline during the late 14th and 15th centuries. The late-medieval decline is also known from different sources, but it is still in many ways enigmatic and interpretations of its character and magnitude are tentative and debatable. One of the most important conclusions presented here is that it is possible to identify the late-medieval decline in pollen diagrams, not only in the investigation area in northern Scania but also in other upland parts of southern Sweden. As it is detectable in pollen diagrams, the decline must have resulted in a significant vegetation change at the time, although not of the same magnitude as the forestation during the late 19th–early 20th century.

Both the medieval regression and the more recent one have been discussed in previous chapters, and I will not repeat that discussion here. Instead my intention is to broaden the perspective and to discuss why regression did not also happen during other periods. In my earlier research, and also at the beginning of this project, I regarded regression and abandonment as something that must have happened during all periods. Even though we may expect abandonment to have been more frequent during periods of large-scale epidemics or societal collapse, there must have been many kinds of difficulties that could lead to abandonment on a more local scale. At farm level a large number of such possible difficulties come to mind: childless farmers may have died of old age, accident or disease with no one to take over, the burden of taxes or rent may have been too heavy, bad weather may have spoiled the harvest, cattle may have died from murrain, etc. Some of the difficulties would have occurred rather randomly in space and time and so would abandonment.

In these circumstances I was surprised by the new results from northern Scania and by the result of the compilation of a larger number of sites (see figure 44b). Neither the Scanian results nor the compilation support the idea of random abandonment during all periods. On the contrary, they give a strong impression of two major regression periods with no or very little regression in the interval in between. There may, of course, have been minor regressions, which have not been captured by the pollen diagrams, but they must have been short-lasting (shorter than a century) and any reforestation that they may have resulted in must have been very local. So, to be precise, in the temporal and spatial scale studied by pollen analyses, regression during the last millennium has been concentrated to two distinct periods. Why is that?

The explanation is probably to be found in the structure of the agricultural society and particularly in the relationship between population numbers and agricultural land. The two major regression periods, i.e. the medieval decline on the one hand and the changes in the late 19th–20th centuries on the other, were very different in character, but they had one thing in common – both of them were connected to significant population decline. During the later regression period, there was not an overall population decline, rather the opposite, but locally in marginal areas there was strong depopulation due to migration from the countryside to cities and abroad. During the medieval regression the population decline was general and probably affected all areas and all parts of society, but as people may have migrated to central areas with fertile soils, abandonment may also in this case have been most extensive in marginal areas.

Hard times and local crises on individual farms must have occurred now and then during all periods, but if they did not coincide with significant population decline in society, or at least in the local area or region, they did not result in farm abandonment. Even though single fields or meadows may have been taken out of use, the entire farm was not abandoned. It seems as if in agricultural societies under normal conditions, i.e. when there was not a period of population decline, there has always been someone ready to take over a farm in decline. They could be crofters, disinherited sons, or others who saw an opportunity to improve their situation. An important factor in this connection is that much less work is needed to maintain land-use and to keep a farm going than to establish a new one (this has been defined as the maintenance hypothesis by Emanuelsson 2005). For this reason, the agricultural society did not allow farms to be abandoned as long as there was manpower available and people ready to take over. In the other situation, when a crisis on the local farm coincided with a general population decline in the area, the farm could be abandoned.

The explanation can be seen as a threshold model, where the magnitude of population decline had to go beyond a certain value for farm abandonment to occur (figure 47). There may have been local tragedies and small-scale fluctuations, but as long as population numbers did not decline beyond a certain level, the difficulties did not normally lead to abandonment. On the other hand, when population decline went beyond the threshold value and abandonment started, a series of feedback mechanisms may have deepened the crisis. Food production decreased, and due to the abandonment much of the work and time once invested in the establishment of farms and their agricultural land was lost. Not until later, when society and pop-

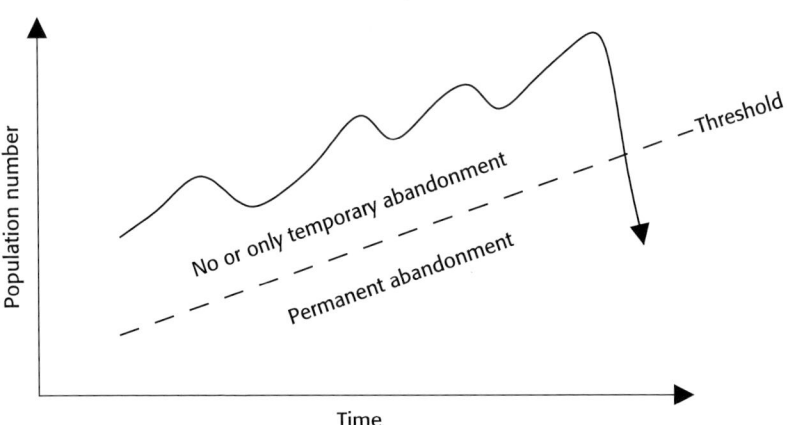

Fig. 47. A threshold model for the relationship between population de-cline and permanent abandonment of settlement and agricultural land.

ulation numbers started to recover, was there a surplus of man-power, which could be put into forest clearing and farm estab-lishment. To, as far as possible, avoid this negative effect of farm abandonment, i.e. the losing of all work investment, there seem to have been at least two different strategies. One, which was practised during the decline, was to keep the land open and to prevent reforestation on the abandoned farm by exten-sive grazing, as in the Värsjö Utmark example (see p. 72). The other, which was practised during the expansion after the cri-sis, was to come back and re-establish arable that had once been cleared from stones, as in the Grisavad example (see p. 95).

The threshold value must have varied between different so-cieties and different agricultural systems. It may be possible to specify it in some cases, but for the Middle Ages in northern Scania we have far too little background information on pop-ulation size and other criteria to do so. The model was used here only to explain in principle why regression and abandon-ment, as reflected in the pollen diagrams, are restricted to two distinct periods. Focus has been on land-use and farms, but the idea of a threshold value and feedback mechanisms when a certain point has been reached may also be used in a more general way. During the Late Middle Ages the decline resulted in a deep crisis with collapsing structures, social conflicts, and riots (Myrdal 1999a). Although it was not a complete societal collapse (cf. definition by Tainter 1988), the crisis may to some degree have resulted in new structures, which in turn was the basis for the expansion that followed during the 16th century (cf. Myrdal 2005). In the investigation area in northern Scania the 16th century was characterized not only by strong agricultural expansion, but also by iron production that was more large-scale and of a more industrial character than before the crisis. It probably reflects reorganization during the decline.

After the strong expansion of the 16th century followed harder times during the 17th century, with war, plunder, and plague. However, the diagrams do not show any abandonment or significant regression during that period (apart for the possible overgrowing of some pastures). If the threshold model is correct, we may conclude that there was no population decline, or at least not a decline that was strong enough to result in permanent farm abandonment or reforestation, during the 17th century, in spite of the obvious difficulties of that period. It seems as if burnt-down farms and devastated agricultural land were soon put into use again. The situation was similar in connection with the Seven Years' War of the late 16th century, which was thoroughly studied by Österberg (1971). She showed that plundered farms received tax reduction, but also that they were soon back as full tax-paying units, usually after less than ten years.

To sum up, regression that was significant enough to lead to abandonment and landscape change – and to be recorded in pollen diagrams – was connected to population decline. Also difficulties during times with positive population trends may have led to some regression, but to a much smaller extent. The question of whether population decline resulted in agricultural regression or *vice versa* is somewhat like the chicken or the egg, and it is not important for the threshold model as such. However,

for the late-medieval regression it is likely that population decline caused by the Black Death was the starting point, but
once started the decline was deepened by feedback mechanisms.

The process of depopulation and regression during the late
19th and early 20th centuries was of a completely different character and does not reflect a general crisis in society. While there
were agricultural regression and reforestation in many marginal areas, central agriculture districts witnessed expansion, and
so did urban areas and industries. The depopulation of the
countryside in the uplands of northern Scania may be explained by the combined effect of push factors, such as population
pressure and poor soils, and pull factors such as vacant jobs in
the cities. The long-lasting effect was a much more polarized
landscape than before, with sharp contrasts between forested
and sparsely populated uplands on the one hand and cultivated and densely populated plains on the other.

New questions

The discussions and interpretations presented in this book
have been based primarily on palaeoecological data, but with
important contributions from archaeology and from the historical record. Many of the conclusions would not have been
possible without such an interdisciplinary approach. It is my
belief that palaeoecology has to be put in an interdisciplinary
context to provide interesting and meaningful interpretations,
in particular when we are dealing with the last millennium.

The investigation has shed light on the landscape and settlement development and many interpretations and suggestions
have been presented regarding landscape change and possible
underlying causes. However, as always, new interpretations
lead to new questions, and I would like to finish here by pointing out some possible directions for future research on Swedish
landscape development.

One important aim would be to quantify. The interpretation of history depends much on quantification and the same is

true for palaeoecology. In spite of worthy attempts to estimate (Andersson Palm 2001), we still know relatively little about medieval population numbers and their regional distribution, and consequently it is difficult to quantify the late-medieval decline and population drop based on the historical record alone (Myrdal 2003). Also the archaeological record is scarce when it comes to investigated rural settlement from the Middle Ages, in particular in forest regions where most palaeoecological research has been carried out (Ersgård & Hållans 1996). Excavations are needed to characterize medieval settlement patterns, social organization, family size, etc. From a palaeoecological point of view, we need to quantify landscape change and landscape openness. In this field of research impressive methodological progress has been made during the last few years, in particular within the Pollen–Landscape Calibration Network (POLLANDCAL), which uses modern analogues, mathematical models and computer simulations to interpret fossil pollen spectra and to estimate past landscape openness (Anderson et al. 2006, Gaillard et al. in press). In a project in progress these methods will be applied to the new pollen data from northern Scania (Broström, Lagerås et al. in progress). It is still complicated, however, and we still need to have an interdisciplinary approach to the study of past landscapes.

Another interesting aim would be to reveal vegetation and land-use dynamics on a finer scale than in the present study. It has recently been put forward that many plant and animal species that we normally regard as typical for the cultural landscape are dependent on abandoned agriculture land to survive in the long run, and that the high biodiversity of the European landscape may in fact partly be a result of re-occurring periods of abandonment and overgrowing (Emanuelsson 2005). The interpretations presented here give some support to this opinion by revealing a dynamic landscape development and by providing details about periods of abandonment. However, in the discussion focus has been at the farm level, i.e. on landscape changes at an approximate scale of a few hectares. To reveal the full spatial complexity of the cultural landscape a finer resolution will be needed, both spatial and temporal. The pollen diagrams of the present study could be used as background for such very local studies of small peat hollows. Also a higher temporal resolution, i.e. denser subsampling of the peat cores, on selected time intervals, would add important information, for instance on vegetation succession after abandonment. A fine scale in palaeoecology is important for its applicability in discussions about nature conservation issues.

Hence, very detailed and local studies would provide new results, but, at the other end of the scale, there is also a need for further regional compilations. To understand the representativeness of local pollen diagrams it is necessary to make comparisons between sites and to combine them into regional syntheses. Sweden has a long tradition of pollen-analytical research and there is a wealth of published pollen diagrams that may be used for compilation. This has also been done by several authors (e.g. Berglund 1969, Berglund et al. 1996b, 2002, Björse & Bradshaw 1998, Holmqvist 2005). However, as the diagrams are of different character and resolution, and their absolute chronologies are of different quality, most compilations have been only on a broad scale. To make more detailed compilations and comparisons it is necessary to evaluate the diagrams and to select only those that are suitable for the task.

These were a few examples of possible directions for future resarch on palaeoecology and landscape change of the last millennium, and, of course, many more could be listed. The study of past landscapes and societies is a never-ending and fascinating research area. Furthermore, it may offer an important long-term perspective to present conditions and decision making, but, significantly, it also offers possibilties for lively stimulating interdisciplinary collaboration and discussion.

REFERENCES

....

Aaby, B. 1994. NAP percentages as an expression of cleared areas. In: Frenzel, B. (ed.) *Evaluation of land surfaces cleared from forests in the Roman Iron Age and the time of migrating Germanic tribes based on regional pollen diagrams*, 13–27. Paläoklimaforschung 12. Gustav Fischer, Stuttgart.

Aaby, B., Digerfeldt, G. 1986. Sampling techniques for lakes and bogs. In: Berglund, B. E. (ed.) *Handbook of Holocene palaeoecology and palaeohydrology*, 181–194. Wiley, Chichester.

Alexandersson, H., Andersson, T. 1995. Nederbörd och åska. In: Raab, B., Vedin, H. (eds) *Klimat, sjöar och vattendrag*, 76–90. Sveriges Nationalatlas, Stockholm.

Almquist-Jacobson, H. 1994. *Interaction of Holocene climate, water balance, vegetation, fire, and cultural land-use in the Swedish borderland*. Lundqua Thesis 30. Department of Quaternary Geology, Lund University.

Andersen, S. T. 1970. *The relative pollen productivity and pollen representation of North European trees, and correction factors for tree pollen spectra, determined by surface pollen analyses from forests*. The Geological Survey of Denmark, Series II 96. Copenhagen.

Andersen, S. T. 1979. Identification of wild grass and cereal pollen. The Geological Survey of Denmark, *Årbog* 1978: 68–92.

Anderson, N. J., Bugmann, H., Dearing, J., Gaillard, M.-J. 2006. Linking palaeoenvironmental data and models to understand the past and to predict the future. *Trends in Ecology and Evolution (TREE)* 21: 696–704.

Andersson Palm, L. *Folkmängden i Sveriges socknar och kommuner 1571–1997: med särskild hänsyn till perioden 1571–1751*. Department of History, Gothenburg University.

Andersson Palm, L. 2001. *Livet, kärleken och döden: fyra uppsatser om svensk befolkningsutveckling 1300–1850*. Department of History, Gothenburg University.

Anthony, D. W. 1990. Migration in archaeology: the baby and the bathwater. *American Anthropologist* 92: 895–914.

Åstrand, J. 2006. *En medeltida skogsgård vid Markaryd: särskild arkeologisk undersökning av RAÄ 75, Markaryds socken, Sm*. Smålands museum rapport 2006:45. Växjö.

Bååth, K. 1983. *Öde sedan stora döden var...: bebyggelse och befolkning i Norra Vedbo under senmedeltid och 1500-tal.* Liber/Gleerup, Lund.

Bartholin, T. S., Berglund, B. E., Malmer, N. 1981. Vegetation and environment in the Gårdlösa area during the Iron Age. In: Stjernquist, B. (ed) *Gårdlösa: an Iron Age community in its natural and social setting. I,* 45–53. Acta Regiae Societatis Humaniorum Litterarum Lundensis LXXXV.

Bartholin, T. S., Berglund, B. E. 1992. The prehistoric landscape in the Köpinge area: a reconstruction based on charcoal analysis. In: Larsson, L., Callmer, J., Stjernquist, B. (eds) *The archaeology of the cultural landscape: field work and research in a south Swedish rural region,* 345–358. Acta Archaeologica Lundensia Series in 4° 19. Almqvist & Wiksell, Stockholm.

Behre, K.-E. 1981. The interpretation of anthropogenic indicators in pollen diagrams. *Pollen et Spores* 23: 225–245.

Berglund, B. E. 1969. Vegetation and human influence in South Scandinavia during Prehistoric time. *Oikos Suppl.* 12: 9–28.

Berglund, B. E. (ed.) 1991. *The cultural landscape during 6000 years in southern Sweden: the Ystad Project.* Ecological Bulletins 41. Copenhagen.

Berglund, B. E., Börjesson, K. (eds) 2002. *Markens minnen: landskap och odlingshistoria på småländska höglandet under 6000 år.* Riksantikvarieämbetet, Stockholm.

Berglund, B. E., Ralska-Jasiewiczowa, M. 1986. Pollen analysis and pollen diagrams. In: Berglund, B. E. (ed.) *Handbook of Holocene palaeoecology and palaeohydrology,* 455–484. Wiley, Chichester.

Berglund, B. E., Birks, H. J. B., Ralska-Jasiewiczowa, M., Wright, H. E. (eds) 1996a. *Palaeoecological events during the last 15000 years: regional syntheses of palaeoecological studies in lakes and mires in Europe.* Wiley, Chichester.

Berglund, B. E., Digerfeldt, G., Engelmark, R., Gaillard, M.-J., Karlsson, S., Miller, U., Risberg, J. 1996b. Sweden. In: Berglund, B. E., Birks, H. J. B., Ralska-Jasiewiczowa, M., Wright, H. E. (eds) *Palaeoecological events during the last 15000 years: regional syntheses of palaeoecological studies in lakes and mires in Europe,* 233–280. Wiley, Chichester.

Berglund, B. E., Lagerås, P., Regnéll, J. 2002. Odlingslandskapets historia i Sydsverige: en pollenanalytisk syntes. In: Berglund, B. E., Börjesson, K. (eds) *Markens minnen: landskap och odlingshistoria på småländska höglandet under 6000 år,* 153–174. Riksantikvarieämbetet, Stockholm.

Beug, H.-J. 1961. *Leitfaden der Pollenbestimmung für Mitteleuropa und angrenzende Gebiete.* Lief. 1. Gustav Fischer, Stuttgart.

Birks, H. J. B., Birks, H. H. 1980. *Quaternary palaeoecology.* Edward Arnold, London.

Björkman, L. 1996a. The late Holocene history of beech Fagus sylvatica and Norway spruce Picea abies at stand-scale in southern Sweden. Lundqua Thesis 39. Department of Quaternary Geology, Lund University.

Björkman, L. 1996b. Long-term population dynamics of Fagus sylvatica at the northern limits of its distribution in southern Sweden: a palaeoecological study. *The Holocene* 6: 225–234.

Björkman, L. 1997a. The history of Fagus forest in southwestern Sweden during the last 1500 years. *The Holocene* 7: 419–432.

Björkman, L. 1997b. The role of human disturbance in the Late Holocene establishment of Fagus and Picea forests at Flahult, western Småland, southern Sweden. *Vegetation History and Archaeobotany* 6: 79–90.

Björkman, L. 2000a. *Pollenanalytisk undersökning av en torvmarkslagerföljd från Trälhultet i Biskopstorps naturreservat, Halmstads kommun.* Lundqua Uppdrag 29. Department of Quaternary Geology, Lund University.

Björkman, L. 2000b. *Pollenanalys av en lagerföljd från Uddared, Laholms kommun.* Lundqua Uppdrag 31. Department of Quaternary Geology, Lund University.

Björkman, L. 2001. The role of human disturbance in Late Holocene vegetation changes on Kullaberg, southern Sweden. *Vegetation History and Archaeobotany* 10: 201–210.

Björkman, L. 2003. *Pollenanalytisk undersökning av en torvmarkslagerföljd från den arkeologiska undersökningslokalen "Område 2" nordost om Köphult inför ombyggnaden av E4: an, delen länsgränsen till Strömsnäsbruk, Markaryds kommun.* Lundqua Uppdrag 48. Department of Quaternary Geology, Lund University.

Björkman, L. 2005. *Pollenanalytisk undersökning av en torvmarkslagerföljd från Baggabygget i Rönnö naturreservat, Laholms kommun.* Lundqua Uppdrag 55. Department of Quaternary Geology, Lund University.

Björkman, L., Bradshaw, R. 1996. The immigration of Fagus sylvatica L. and Picea abies (L.) Karst. into a natural forest stand in southern Sweden during the last 2000 years. *Journal of Biogeography* 23: 235–244.

Björkman, L., Ekström, J. 2002. *Paleoekologisk förundersökning av torvmarker inför ombyggnaden av E4:an sträckan Örkelljunga–länsgränsen i nordvästra Skåne.* Lundqua Uppdrag 39. Department of Quaternary Geology, Lund University.

Björkman, L., Ekström, J. 2003. *Pollenanalytisk undersökning av en torvmarkslagerföljd från den arkeologiska undersökningslokalen "Område 12/13" nordväst om Exhult inför ombyggnaden av E4:an, delen länsgränsen till Strömsnäsbruk, Markaryds kommun.* Lundqua Uppdrag 47. Department of Quaternary Geology, Lund University.

Björkman, L., Persson, T. 2005. *Pollenanalytisk undersökning av en torvmarkslagerföljd från Käringsjön i Övraby socken, Halmstad kommun.* Lundqua Uppdrag 56. Department of Quaternary Geology, Lund University.

Björse, G., Bradshaw, R. 1998. 2000 years of forest dynamics in southern Sweden: suggestions for forest management. *Forest Ecology and Management* 104: 15–26.

Bjuggner, E., Rosengren, E. 1999. Likt eller olikt?: bebyggelse på den sydhalländska landsbygden under tidig medeltid. In: Artelius, T., Englund, E., Ersgård, L. (eds) *Kring västsvenska hus: boendets organisation och symbolik i förhistorisk och historisk tid*, 87–97. Gotarc Serie C. Arkeologiska skrifter 22. Department of Archaeology, Gothenburg University.

Blennow, K., Bärring, L., Jönsson, P., Linderson, M.-L., Mattson, J. O., Schlyter, P. 1999. Klimat, sjöar och vattendrag. In: Germundsson, T., Schlyter, P. (eds) *Atlas över Skåne*, 30–37. Sveriges Nationalatlas, Stockholm.

Boserup, E. 1965. *The conditions of agricultural growth: the economics of agrarian change under population pressure.* Allen & Unwin, London.

Bradshaw, R. H. W. 1988. Spatially-precise studies of forest dynamics. In: Huntley, B., Webb, T. III (eds) *Vegetation history*, 725–751. Kluwer, Dordrecht.

Bradshaw, R. H. W., Hannon, G. 1992. Climatic change, human influence and disturbance regime in the control of vegetation dynamics within Fiby Forest, Sweden. *Journal of Ecology* 80: 625–632.

Bradshaw, R. H. W., Coxon, P., Greig, J. R. A., Hall, A. R. 1981. New fossil evidence for the past cultivation and processing of hemp (Cannabis sativa L.) in eastern England. *New Phytologist* 89: 503–510.

Bradshaw, R. H. W., Holmqvist, B. H., Cowling, S. A., Sykes, M. T. 2000. The effects of climate change on the distribution and management of Picea abies in southern Scandinavia. *Canadian Journal of Forest Research* 30: 1992–1998.

Bringéus, N.-A. 1983. *Resekörare och besekrämare: om handeln i Örkelljungabygden i gången tid.* Signum, Lund.

Broström, A., Gaillard, M.-J., Ihse, M., Odgaard, B. 1998. Pollen-landscape relationships in modern analogues of ancient cultural landscapes in southern Sweden: a first step towards quantification of vegetation openness in the past. *Vegetation History and Archaeobotany* 7: 189–201.

Broström, A., Sugita, S., Gaillard, M.-J. 2004. Pollen productivity estimates for the reconstruction of past vegetation cover in the cultural landscape of southern Sweden. *The Holocene* 14: 368–381.

Campbell, B. M. S. 1995. Ecology versus economics in late Thirteenth- and early Fourteenth-Century English agriculture. In: Sweeney, D. (ed.) *Agriculture in the Middle Ages: technology, practice, and representation,* 76–108. University of Pennsylvania Press, Philadelphia.

Carter, T. R., Parry, M. L. 1984. Strategies for assessing impacts of climate change in marginal areas. In: Mörner, N.-A., Karlén, W. (eds) *Climatic changes on a yearly to millennial basis,* 401–412. Reidel, Dordrecht.

Cohn, S. K. Jr. 2003. *The Black Death transformed: disease and culture in early Renaissance Europe.* Arnold, London.

Connelid, P. 2002. Åker, toft och vång: landskapsförändringar i skånsk skogsbygd från vikingatid till cirka 1800. In: Carlie, A. (ed.) *Skånska regioner: tusen år av kultur och samhälle i förändring,* 413–467. Arkeologiska undersökningar, Skrifter 40. Riksantikvarieämbetet, Stockholm.

Connelid, P. 2003. Fossila bandparceller i nordvästra Skåne. In: Olsson, M. (ed.) *Kartering och omlandsstudier utmed E4:an i norra Skåne. Arkeologiska utredningar steg 2a för väg E4, Örkelljunga–länsgränsen och riksväg 24, Bälinge–Västra Spång.* Riksantikvarieämbetet, Avdelningen för arkeologiska undersökningar, UV Syd Rapport 2003:19.

Connelid, P. 2004. Byarna kring Falkenberg: jordbrukslandskapet i Stafsinge, Tröinge och Skrea från vikingatiden till ca 1800. In: Carlie, L., Ryberg, E., Streiffert, J., Wranning, P. (eds) *Landskap i förändring–hållplatser i det förgångna: artiklar med avstamp i de arkeologiska undersökningarna för Västkustbanans dubbelspår förbi Falkenberg i Halland,* 359–398. Halland Länsmuseer, Halmstad/Riksantikvarieämbetet, Stockholm.

Connelid, P., Mascher, C., Weiler, E. 1993. Röstorp: ett västsvenskt röjningsröseområde i skogsmark. In: *Arkeologi i Sverige* (New series) 2: 15–38. Riksantikvarieämbetet, Stockholm.

Dahl, S. 1942. *Torna och Bara: studier i Skånes bebyggelse- och näringsgeografi före 1860*. Meddelanden från Lunds Universitets Geografiska Institution, Avhandlingar 6. Lund.

Digerfeldt, G. 1982. *The Holocene development of Lake Sämbosjön. 1 The regional vegetation history*. Lundqua Report 23. Department of Quaternary Geology, Lund University.

Duby, G. 1973. *Krigare och bönder: den europeiska ekonomins första uppsving 600–1200*. Norstedt, Stockholm.

Edlund, L. 2004. Skånska våtmarkers historia: utveckling av en metodik för att med hjälp av lagerföljdsdata och Geografiska informationssystem (GIS) modellera och rekonstruera våtmarksförändringar. Unpublished report.

Edwards, K. J. 1991. Using space in cultural palynology: the value of the off-site pollen record. In: Harris, D. R., Thomas, K. D. (eds) *Modelling ecological change: perspectives from neoecology, palaeoecology and environmental archaeology*, 61–73. Institute of Archaeology, University College London.

Edwards, K. J., Whittington, G. 1992. Male and female plant selection in the cultivation of hemp, and variations in fossil Cannabis pollen representation. *The Holocene* 2: 85–87.

Ekstam, U., Aronsson, M., Forshed, N. 1988. *Ängar: om naturliga slåttermarker i odlingslandskapet*. LTs förlag, Stockholm.

Ekström, J. 2000. *Pollenanalys av en torvlagerföljd från Rydholmskärret: en miljöarkeologisk undersökning inför ombyggnad av väg 897 sträckan Sandsbro–Stockekvarn, Gårdsby socken, Växjö kommun*. Lundqua Uppdrag 27. Department of Quaternary Geology, Lund University.

Eliasson, P. 2002. *Skog, makt och människor: en miljöhistoria om svensk skog 1800–1875*. Skogs- och lantbruksakademiens meddelanden 25. Kungl. Skogs- och Lantbruksakademien, Stockholm.

Emanuelsson, M. 2001. *Settlement and land-use history in the central Swedish forest region: the use of pollen analysis in interdisciplinary studies*. Silvestria 223. Swedish Univeristy of Agricultural Sciences.

Emanuelsson, M., Johansson, A., Nilsson, S., Pettersson, S., Svensson, E. 2003. *Settlement, shieling and landscape: the local history of a forest hamlet*. Lund Studies in Medieval Archaeology 32. Almqvist & Wiksell, Stockholm.

Emanuelsson, U. 1988. A model for describing the development of the cultural landscape. In: Birks, H. H., Birks, H. J. B., Kaland, P. E., Moe, D. (eds) *The cultural landscape: past, present and future*, 111–121. Cambridge University Press, Cambridge.

Emanuelsson, U. 2005. Ohävd – en nödvändig hävd. In: *Arkeologi och naturvetenskap*, 111–128. Symposier på Krapperups borg 6. Gyllenstiernska Krapperupstiftelsen, Nyhamnsläge.

Engelmark, R. 1992. A review of the farming economy in South Scania based on botanical evidence. In: Larsson, L., Callmer, J., Stjernquist, B. (eds) *The archaeology of the cultural landscape: field work and research in a south Swedish rural region*, 369–375. Acta Archaeologica Lundensia Series in 4° 19. Almqvist & Wiksell, Stockholm.

Engelmark, R., Viklund, K. 1990. Makrofossilanalys av växter: kunskap om odlandets karaktär och historia. *Bebyggelsehistorisk tidskrift* 19: 33–41.

Englund, L.-E. 1995. Skånska och halländska medeltida blästor: nya aspekter på gamla problem. In: Olsson, S.-O. (ed.) *Medeltida danskt järn: framställning av och handel med järn i Skåneland och Småland under medeltiden*, 79–92. Forskning i Halmstad 1. Centrum för Sydsvensk Kulturmiljöforskning, Halmstad.

Englund, L.-E. 2002. *Blästbruk: myrjärnshanteringens förändringar i ett långtidsperspektiv*. Jernkontorets Bergshistoriska Skriftserie 40. Stockholm.

Erdtman, G. 1960. The acetolysis method. *Svensk Botanisk Tidskrift* 54: 561–564.

Ericson, A. 2004. Medeltida odlingar på utmarker: krisfenomen eller överskottsproduktion? *Tidskrift* 4: 41–70.

Ersgård, L., Hållans, A.-M. 1996. *Medeltida landsbygd: en arkeologisk utvärdering*. Arkeologiska undersökningar, Skrifter 15. Riksantikvarieämbetet, Stockholm.

Fagan, B. 2000. *The Little Ice Age: how climate made history 1300–1850*. Basic Books, New York.

Forenius, S., Willim, A., Grandin, L. 2005. Medeltida blästbruk vid Bredabäck: E4-projektet i Skåne, område E4:31. Riksantikvarieämbetet, unpublished report.

Fægri, K., Iversen, J. 1989. *Textbook of Pollen Analysis* (4th edition, revised by Fægri, K., Kaland, P. E., Krzywinski, K.). Wiley, Chichester.

Gadd, C.-J. 2000. *Den agrara revolutionen 1700–1870*. Det svenska jordbrukets historia. Natur och Kultur/LT, Stockholm.

Gaillard, M.-J., Göransson, H. 1991. The Bjäresjö area: vegetation and landscape through time. In: Berglund, B. E. (ed.) *The cultural landscape during 6000 years in southern Sweden: the Ystad Project*, 167–174. Ecological Bulletins 41. Copenhagen.

Gaillard, M.-J., Birks, H. J. B., Emanuelsson, U., Berglund, B. E. 1992. Modern pollen/land-use relationships as an aid in the reconstruction of past land-uses and cultural landscapes: an example from south Sweden. *Vegetation History and Archaeobotany* 1: 3–17.

Gaillard, M.-J., Birks, H. J. B., Emanuelsson, U., Karlsson, S., Lagerås, P., Olausson, D. 1994. Application of modern pollen/land-use relationships to the interpretation of pollen diagrams: reconstruction of land-use history in south Sweden. *Review of Palaeobotany and Palynology* 82: 47–73.

Gaillard, M.-J., Hannon, G. E., Håkansson, H., Olsson, S., Possnert, G., Sandgren, P. 1996. New data on the Holocene forest and land-use history of Skåne based on AMS ^{14}C dates of terrestrial plant macroremains, and biostratigraphical, chemical, and mineral magnetic analyses of lake sediments. *GFF* 118A: 65–66.

Gaillard, M.-J., Sugita, S., Bunting, M.-J., Middleton, D., Hicks, S., Broström, A., Hellman, S., Caseldine, C., Giesecke, T., Hjelle, K., Langdon, C. Nielsen, A.-B., Poska, A., von Stedingk, H., Veski, S. and POLLANDCAL members in press. The use of simulation models in reconstructing past landscapes from fossil pollen data: research strategy and results from the POLLANDCAL network. *Vegetation History and Archaeobotany.*

Gauffin, S. 1981. *Ödesbölet Svedäng, Alsens socken: rapport från en arkeologisk undersökning.* Jämtlands läns museum, Östersund.

Giesecke, T., Bennett, K. D. 2004. The Holocene spread of Picea abies (L.) Karst. in Fennoscandia and adjacent areas. *Journal of Biogeography* 31: 1523–1548.

Gissel, S., Jutikkala, E., Österberg, E., Sandnes, J., Teitson, B. (eds) 1981. *Desertion and land colonization in the Nordic countries c. 1300–1600: comparative report from the Scandinavian research project on deserted farms and villages.* Almqvist & Wiksell, Stockholm.

Gräslund, B. 1979. Efterskrift. In: Stenberger, M. *Det forntida Sverige*, 871–912. Almqvist & Wiksell, Stockholm.

Greig, J. 1983. The palaeoecology of some British hay meadow types. In: van Zeist, W., Casparie, W. A. (eds) *Plants and ancient man: studies in palaeoethnobotany*, 213–226. A. A. Balkema, Rotterdam.

Gren, L. 1989. Det småländska höglandets röjningsröseområden. In: *Arkeologi i Sverige 1986*, 73–95. Riksantikvarieämbetet, Stockholm.

Grove, J. M. 2001. The onset of the Little Ice Age. In: Jones, P. D., Ogilvie, A. E. J., Davies, T. D., Briffa, K. R. (eds) *History and climate: memories of the future?*, 153–185. Kluwer, New York.

Haaland, S. 2002. *Fem tusen år med flammer: det europeiske lyngheilandskapet.* Vigmostad & Bjørke, Bergen.

Hall, V. 2003. Vegetation history of mid- to western Ireland in the 2nd millenium A.D.; fresh evidence from tephra-dated palynological investigations. *Vegetation History and Archaeobotany* 12: 7–17.

Hansson, A.-M. 1996. Finds of hops, Humulus lupulus L., in the Black Earth of Birka, Sweden. In: *Proceedings from the 6th Nordic Conference on the Application of Scientific Methods in Archaeology, Esbjerg 1993*, 129–137. Arkæologiske rapporter 1. Esbjerg Museum.

Hansson, A., Olsson, C., Storå, J., Welinder, S., Zetterström, Å. 2005. *Agrarkris och ödegårdar i Jämtland.* Jamtli förlag, Östersund.

Hatcher, J. 1977. *Plague, population and the English economy 1348–1530.* Basingstoke, London.

Heckscher, E. F. 1957. *Svenskt arbete och liv: från medeltiden till nutiden.* Aldus/Bonniers, Stockholm.

Hedberg, H. D. (ed.) 1976. *International stratigraphic guide: a guide to stratigraphic classification, terminology, and procedure.* Wiley, Chichester.

Hjelle, K. L. 1999. Modern pollen assemblages from mown and grazed vegetation types in western Norway. *Review of Palaeobotany and Palynology* 107: 55–81.

Hjelmqvist, H. 1992. Some economic plants from the prehistoric and mediaeval periods in southern Sweden. In: Larsson, L., Callmer, J., Stjernquist, B. (eds) *The archaeology of the cultural landscape: field work and research in a south Swedish rural region*, 359–367. Acta Archaeologica Lundensia Series in 4° 19. Almqvist & Wiksell, Stockholm.

Holmqvist, B. H. 2005. *Classification of large pollen datasets using neural networks with application and modelling pollen data.* Lundqua Report 39. Department of Quaternary Geology, Lund University.

Hybel, N. 1989. *Crisis or change: the concept of crisis in the light of agrarian structural reorganization in late medieval England*. Aarhus University Press, Aarhus.

Hyenstrand, Å. 1979. *Ancient monuments and prehistoric society*. Riksantikvarieämbetet, Stockholm.

Jacobson, G. L., Bradshaw, R. H. W. 1981. The selection of sites for paleovegetational studies. *Quaternary Research* 16: 80–96.

Joelsson, K. 2006. *De halländska ljungmarkerna och deras försvinnande: en agrarhistorisk studie*. Examensarbete i agrarhistoria. Institutionen för ekonomi, Sveriges Lantbruksuniversitet. Examensarbete 435. (Examination paper in agrarian history from the Swedish University of Agricultural Sciences.)

Jóhansen, J. 1996. Faroe Islands. In: Berglund, B. E., Birks, H. J. B., Ralska-Jasiewiczowa, M., Wright, H. E. (ed.) *Palaeoecological events during the last 15 000 years: regional syntheses of palaeoecological studies of lakes and mires in Europe*, 145–152. Wiley, Chichester.

Jowsey, P. C. 1966. An improved peat sampler. *New Phytologist* 65: 245–248.

Karlsson, S. 2000. Kan medeltida järnhantering i norra Skåne spåras med hjälp av pollenanalys. In: Ödman, A. (ed.) *Järn: Wittsjökonferensen 1999*, 125–149. Norra Skånes medeltid 1. Report Series 75. Institute of Archaeology, Lund University.

Kershaw, I. 1973. The great famine and agrarian crisis in England 1315–1322. *Past and Present* 59: 3–50.

Knarrström, A. 2003. E4:16: sentida bebyggelse- och odlingslämningar vid torpet Grisavad. In: Jacobsson, B. (ed.) *Förundersökningar utmed E4:an i norra Skåne*, 29–41. Riksantikvarieämbetet UV Syd Rapport 2003:21.

Knarrström, A. 2004. En sentida skogsgård vid Värsjö: kol, tjära och odling. Område E4:34A, Örkelljunga–länsgränsen. Riksantikvarieämbetet UV Syd, Dokumentation av fältarbetsfasen 2004:4. (Unpublished report)

Knarrström. B. (ed.) 2007. *Stenåldersjägare*. Riksantikvarieämbetet, Stockholm.

Königsson, L.-K. 1989. Human impact trends in the landscape development at Hjärtenholm during the last 5000 years. *Striae* 25: 59–73.

Königsson, L.-K-, Qvarfort, U. 1988. Den förhistoriska järnframställningen på Åsamon i Tabergs Berggslag. *Tabergs Bergslag* 15: 49–69.

REFERENCES

Krok, T. O. B. N., Almquist, S. 1994. *Svensk flora: fanerogamer och ormbunksväxter* (26th edition revised by Jonsell, L., Jonsell, B.). Esselte, Uppsala.

Küster, H. 1997. The role of farming in the postglacial expansion of beech and hornbeam in the oak woodlands of central Europe. *The Holocene* 7: 239–242.

Lagerås, P. 1996a. *Vegetation and land-use in the Småland Uplands, southern Sweden, during the last 6000 years.* Lundqua Thesis 36. Department of Quaternary Geology, Lund University.

Lagerås, P. 1996b. Long-term history of land-use and vegetation at Femtingagölen: a small lake in the Småland Uplands, southern Sweden. *Vegetation History and Archaeobotany* 5: 215–228.

Lagerås, P. 1996c. Farming and forest dynamics in an agriculturally marginal area of southern Sweden, 5000 BC to present: a palynological study of Lake Avegöl. *The Holocene* 6: 301–314.

Lagerås, P. 1997. Våtmarker i väglinjen: rekognoscering av torvmarker och sjöar. In: Wallin, L., Olsson, M., Connelid, P., Karsten, P., Knarrström, B., Lagerås, P., Mattisson, A., Olsson, M., Skansjö, S. *Arkeologisk utredning från Örkelljunga till länsgränsen. Särskild arkeologisk utredning steg 1, väg E4, förbi Örkelljunga (Eket–Värsjö) och förbi Fagerhult (Värsjö–Köphult) samt väg 24, delen Bälinge–Västra Spång, Skåne,* 90–99. Riksantikvarieämbetet, UV Syd Rapport 1997:58.

Lagerås, P. (ed.) 2000a. *Arkeologi och paleoekologi i sydvästra Småland: tio artiklar från Hamnedaprojektet.* Arkeologiska undersökningar, Skrifter 34. Riksantikvarieämbetet, Stockholm.

Lagerås, P. 2000b. Järnålderns odlingssystem och landskapets långsiktiga förändring: Hamnedas röjningsröseområden i ett paleoekologiskt perspektiv. In: Lagerås, P. (ed.) *Arkeologi och paleoekologi i sydvästra Småland: tio artiklar från Hamnedaprojektet,* 167–229. Arkeologiska undersökningar, Skrifter 34. Riksantikvarieämbetet, Stockholm

Lagerås, P. 2002a. Röjningsrösen och den historiska bygden: brukandet av till synes ålderdomliga röseområden under historisk tid. *Tidskrift* 2: 25–43.

Lagerås, P. 2002b. Skog, slåtter och stenröjning: paleoekologiska undersökningar i trakten av Stoby i norra Skåne. In: Carlie, A. (ed.) *Skånska regioner: tusen år av kultur och samhälle i förändring,* 363–411. Arkeologiska undersökningar, Skrifter 40. Riksantikvarieämbetet, Stockholm.

Lagerås, P. 2003a. Approaches and methods for commissioned archaeology in wetlands: experience from the E4 Project in Skåne, southern Sweden. *European Journal of Archaeology* 6: 231–249.

Lagerås, P. 2003b. Torvstratigrafier. In: Olsson, M. (ed.) *Kartering och omlandsstudier utmed E4:an i norra Skåne. Arkeologiska utredningar steg 2a för väg E4, Örkelljunga–länsgränsen och riksväg 24, BälingeVästra Spång*, 162–217. Riksantikvarieämbetet, UV Syd Rapport 2003:19.

Lagerås, P., Bartholin, T. S. 2003. Fire and stone clearance in Iron Age agriculture: new insights inferred from the analysis of terrestrial macroscopic charcoal in clearance cairns in Hamneda, southern Sweden. *Vegetation History and Archaeobotany* 12: 83–92.

Lagerås, P., Jansson, K., Vestbö, A. 1995. Land-use history of the Axlarp area in the Småland uplands, southern Sweden: palaeoecological and archaeological investigations. *Vegetation History and Archaeobotany* 4: 223–234.

Lagerås, P., Sandgren, P. 1994. The use of mineral magnetic analyses in identifying Middle and Late Holocene agriculture: a study of peat profiles in Småland, southern Sweden. *Journal of Archaeological Science* 21: 687–697.

Lagerås, P., Olsson, M., Wallin, L. 2000. Röjningsrösens utseende och ålder: resultat från E4-projektet i norra Skåne. In: Ersgård, L. (ed.) *Människors platser: tretton arkeologiska studier från UV*, 167–184. Arkeologiska undersökningar, Skrifter 31. Riksantikvarieämbetet, Stockholm.

Lamb, H. H. 1995. *Climate history and the modern world* (2nd edition). Routledge, London.

Larsson, L.-O. 1964. *Det medeltida Värend: studier i det småländska gränslandets historia fram till 1500-talets mitt*. Bibliotheca Historica Lundensis 12. Lund.

Larsson, L-O. 1970. Kronans jordeböcker från 1500-talet och den senmedeltida ödegårdsprocessen: några synpunkter på terminologi och retrospektiv metod. *Historisk tidskrift* 1970 (1): 24–76.

Larsson, L.-O. 1972. *Kolonisation och befolkningsutveckling i det svenska agrarsamhället 1500–1640*. Bibliotheca Historica Lundensis 27. Gleerups, Lund.

Lind, H., Svensson, E., Hansson, J. 2001. *Projekt uppdragsarkeologi: sentida bebyggelse i antikvarisk och arkeologisk verksamhet*. Rapport 2001:2. Riksantikvarieämbetet, Stockholm.

Lindbladh, M. 1998. *Long term dynamics and human influence in the forest landscape of southern Sweden.* Silvestria 78. Swedish University of Agricultural Sciences.

Lindbladh, M., Bradshaw, R. 1995. The development and demise of a medieval forest-meadow system at Linnaeus' birthplace in southern Sweden: implications for conservation and forest history. *Vegetation History and Archaeobotany* 4: 153–160.

Lindbladh, M., Bradshaw, R. 1998. The origin of present forest composition and pattern in southern Sweden. *Journal of Biogeography* 25: 463–477.

Lindman, G. (ed.) 2004. *Gårdar från förr: nordbohusländsk bebyggelsehistoria utifrån arkeologiska undersökningar av tre medeltida gårdar.* Arkeologiska undersökningar, Skrifter 56. Riksantikvarieämbetet, Stockholm.

Linderoth, T. 2004. Odlingslämningar och en neolitisk boplats vid Skånes Fagerhult. Område E4:45, Örkelljunga–länsgränsen. Riksantikvarieämbetet, Avdelningen för arkeologiska undersökningar, DAFF 2004:07. Unpublished report.

Lindkvist, T. 2003. Från träl till landbo: uppkomsten av det medeltida godssystemet i Europa och Norden. In: Lindkvist, T., Myrdal, J. (eds) *Trälar: ofria i agrarsamhället från vikingatid till medeltid,* 9–21. Skrifter om skogs- och lantbrukshistoria 17. Nordiska museet, Stockholm.

Lindquist, B. 1959. Forest vegetation belts in southern Scandinavia. *Acta Hort Gotoburgensis* 22: 111–144.

Linnæus, C. 1751 [reprint 1975]. *Skånska resa.* Wahlström & Widstrand, Stockholm.

Livi Bacci, M. 2000. *The population of Europe: a history.* Blackwell, Oxford.

Ljung, K. 2003. *Pollenanalytisk undersökning av en torvmarkslagerföljd från Lärkesholm, Örkelljunga kommun.* Lundqua Uppdrag 43. Department of Quaternary Geology, Lund University.

Luterbacher, J. 2001. The Late Maunder Minimum (1675–1715): climax of the 'Little Ice Age' in Europe. In: Jones, P. D., Ogilvie, A. E. J., Davies, T. D., Briffa, K. R. (eds) *History and climate: memories of the future?,* 29–54. Kluwer, New York.

Malmer, N. 1965. The south-western dwarf shrub heaths. *Acta Phytogeographica Suecica* 50: 123–130.

Malmer, N. 1968. Om ljunghedar och andra rishedar i Sydvästsverige. *Sveriges Natur Årsbok* 1968: 177–187.

Malmström, C. 1939. Hallands skogar under de senaste 300 åren. *Meddelanden från Statens Skogsförsöksanstalt* 31: 171–300.

Mascher, C. 1992. Västsveriges skogsbygder: ett område med storskalig förhistorisk odling. *Bebyggelsehistorisk tidskrift* 23: 7–26.

Mercuri, A. M., Accorsi, C. A., Mazzanti, M. B. 2002. The long history of Cannabis and its cultivation by the Romans in central Italy, shown by pollen records from Lago Albano and Lago di Nemi. *Vegetation History and Archaeobotany* 11: 263–276.

Moberg, A., Sonechkin, D. M., Holmgren, K., Datsenko, N. M., Karlén, W. 2005. Highly variable Northern Hemisphere temperatures reconstructed from low- and high-resolution proxy data. *Nature* 433: 613–617.

Moore, P. D., Webb, J. A., Collinson, M. E. 1991. *Pollen analysis* (2nd edition). Blackwell, Oxford.

Morell, M. 2001. *Jordbruket i industrisamhället 1870–1945*. Det svenska jordbrukets historia. Natur och Kultur/LT, Stockholm.

Morris, R. 1989. *Churches in the landscape*. Dent, London.

Myrdal, J. 1982. Jordbruksredskap av järn före år 1000. *Fornvännen* 77: 81–104.

Myrdal, J. 1985. *Medeltidens åkerbruk: agrarteknik i Sverige ca 1000 till 1520*. Nordiska museets Handlingar 105. Nordiska museet, Stockholm.

Myrdal, J. 1997. The agricultural transformation of Sweden, 1000–1300. In: Astill, G., Langdon, J. (eds) *Medieval farming and technology: the impact of agricultural change in Northwest Europe*, 147–171. Technology and change in history 1. Brill, Leiden.

Myrdal, J. 1999a. *Jordbruket under feodalismen 1000–1700*. Det svenska jordbrukets historia. Natur och Kultur/LT, Stockholm.

Myrdal, J. 1999b. The agrarian revolution restrained: Swedish agrarian technology in the 16th century in a European perspective. In: Liljewall, B. (ed.) *Agrarian systems in early Modern Europe: technology, tools, trade*, 96–145. Skrifter om skogs- och lantbrukshistoria 13. Nordiska museet, Stockholm.

Myrdal, J. 2003. *Digerdöden, pestvågor och ödeläggelse: ett perspektiv på senmedeltidens Sverige*. Runica et Mediævalia, Stockholm.

Myrdal, J. 2005. Krisen är ingen katastrof. In: *Arkeologi och naturvetenskap*, 82–110. Gyllenstiernska Krapperupstiftelsen, Nyhamnsläge.

Myrdal, J., Tollin, C. 2003. Brytar och tidigmedeltida huvudgårdar. In: Lindkvist, T., Myrdal, J. (eds) *Trälar: ofria i agrarsamhället från vikingatid till medeltid*, 133–168. Skrifter om skogs- och lantbrukshistoria 17. Nordiska museet, Stockholm.

Myrdal, J. 2006. The forgotten plague: the Black Death in Sweden. In: Hämäläinen, P. (ed.) When disease makes history: epidemics and great historical turning points, 141-186. Helsinki University Press, Helsinki.

Nordberg, M. 1995. *I kung Magnus tid: Norden under Magnus Eriksson 1317–1374.* Norstedt, Stockholm

Nordström, O. 1996. Utmarkens invånare: en moderniseringsresurs. In: Liljewall, B. (ed.) *Tjära, barkbröd och vildhonung: utmarkens människor och mångsidiga resurser,* 156–171. Skrifter om skogs- och lantbrukshistoria 9. Nordiska museet, Stockholm.

Odgard, B. V. 1994. *The Holocene vegetation history of northern West Jutland, Denmark.* Opera Botanica 123. Copenhagen.

Ödman, A, 1995. Skånes borgar. In: Mogren, M., Wienberg, J. (eds) *Lindholmen: medeltida riksborg i Skåne,* 31–44. Lund Studies in Medieval Archaeology 17. Almqvist & Wiksell, Stockholm.

Ödman, A. 2001. *Vittsjö – en socken i dansk järnbruksbygd.* Institute of Archaeology, University of Lund, Report Series 76. Lund.

Olsson, I. U. 1986. Radiometric dating. In: Berglund, B. E. (ed.) *Handbook of Holocene palaeoecology and palaeohydrology,* 273–312. Wiley, Chichester.

Olsson, I. U. 1991. Accuracy and precision in sediment chronology. *Hydrobiologia* 214: 25–34.

Österberg, E. 1971. *Gränsbygd under krig: ekonomiska, demografiska och administrativa förhållanden i sydvästra Sverige under och efter nordiska sjuårskriget.* Bibliotheca Historica Lundensis XXVI. Gleerup, Lund.

Österberg, E. 1977. *Kolonisation och kriser: bebyggelse, skattetryck, odling och agrarstruktur i västra Värmland ca 1300–1600.* Bibliotheca Historica Lundensis XLIII. Gleerups, Lund.

Österberg, E. 1981. Methods, hypotheses and study areas. In: Gissel, S., Jutikkala, E., Österberg, E., Sandnes, J., Teitson, B. (eds) *Desertion and land colonization in the Nordic countries c. 1300–1600: comparative report from the Scandinavian research project on deserted farms and villages,* 26–77. Almqvist & Wiksell, Stockholm.

Pamp, B. 1983. *Ortnamn i Skåne.* Almqvist & Wiksell, Stockholm.

Pamp, B. 1988. *Skånska orter och ord.* Corona, Malmö.

Parry, M. L. 1981. Climatic change and the agricultural frontier: a research strategy. In: Wigley, T. M. L., Ingram, M. J., Farmer, G. (eds) *Climate and history: studies in past climates and their impact on Man*, 319–336. Cambridge University Press, Cambridge.

Pedersen, E. A., Widgren, M. 1998. Järnålder 500 f.Kr.–1000 e.Kr. In: Welinder, S., Pedersen, E. A., Widgren, M. *Jordbrukets första femtusen år 4000 f.Kr.–1000 e.Kr.*, 235–459. Det svenska jordbrukets historia. Natur och Kultur/LT, Stockholm.

Postan, M. M. 1972. *The medieval economy and society: an economic history of Britain in the Middle Ages*. Weidenfeld & Nicholson, London.

Peglar, S. 1979. A radiocarbon-dated pollen diagram from Loch of Winless, Caithness, North-East Scotland. *New Phytologist* 82: 245–263.

Peglar, S. M., Fritz, S. C., Birks, H. J. B. 1989. Vegetation and land-use history at Diss, Norfolk, UK. *Journal of Ecology* 77: 203-222.

Persson, B. 2001. *Pestens gåta: farsoter i det tidiga 1700-talets Skåne*. Studia Historica Lundensia 5. Nordic Academic Press, Lund.

Petersson, M. 2006. *Djurhållning och betesdrift: djur, människor och landskap i västra Östergötland under yngre bronsålder och äldre järnålder*. Department of Archaeology and Classical Studies, Uppsala University. Riksantikvarieämbetet, Stockholm.

Poulsen, B. 1997. Agricultural technology in medieval Denmark. In: Astill, G., Langdon, J. (eds) *Medieval farming and technology: the impact of agricultural change in Northwest Europe*, 115–145. Technology and change in history 1. Brill, Leiden.

Prentice, I., C. 1988. Records of vegetation in time and space: the principles of pollen analysis. In: Huntley, B., Webb, T. III (eds) *Vegetation history*, 17–42. Kluwer, Dordrecht.

Prøsch-Danielsen, L., Simonsen, A. 2000. *The deforestation patterns and the establishment of the coastal heathland of southwestern Norway*. AmS-Skrifter 15. Museum of Archaeology, Stavanger.

Rasmussen, P. 2005. Mid- to late-Holocene land-use change and lake development at Dallund Sø, Denmark: vegetation and land-use history inferred from pollen data. *The Holocene* 15: 1116–1129.

Rasmussen, P., Anderson, J. 2005. Natural and anthropogenic forcing of aquatic macrophyte development in a shallow Danish lake during the last 7000 years. *Journal of Biogeography* 32: 1993–2005.

Regnéll, J. 1989. *Vegetation and land use during 6000 years: palaeoecology of the cultural landscape at two lake sites in southern Skåne, Sweden.* Lundqua Thesis 27. Department of Quaternary Geology, Lund University.

Regnéll, J. 1992. Preparing pollen concentrates for AMS dating: a methodological study from a hard-water lake in southern Sweden. *Boreas* 21: 373–377.

Regnéll, J., Everitt, E. 1996. Preparative centrifugation: a new method for preparing pollen concentrates suitable for radiocarbon dating by AMS. *Vegetation History and Archaeobotany* 5: 201–205.

Reille, M. 1995. *Pollen et spores d'Europe et d'Afrique du nord: Supplement 1.* Laboratoire de Botanique Historique et Palynologie, Marseille.

Reille, M. 1998. *Pollen et spores d'Europe et d'Afrique du nord: Supplement 2.* Laboratoire de Botanique Historique et Palynologie, Marseille.

Reille, M. 1999. *Pollen et spores d'Europe et d'Afrique du nord* (2nd edition). Laboratoire de Botanique Historique et Palynologie, Marseille.

Reimer, P. J., Baillie, M. G. L., Bard, E., Bayliss, A., Beck, J W., Bertrand, C.J.H., Blackwell, P. G., Buck, C. E., Burr, G. S., Cutler, K. B., Damon, P. E., Edwards, R L., Fairbanks, R. G., Friedrich, M., Guilderson, T. P., Hogg, A. G., Hughen, K. A., Kromer, B., McCormac, G., Manning, S., Ramsey, C. B., Reimer, R. W., Remmele, S., Southon, J. R., Stuiver, M., Talamo, S., Taylor, F.W., van der Plicht, J., Weyhenmeyer, C. E. 2004. IntCal04 Terrestrial Radiocarbon Age Calibration, 0–26 Cal Kyr BP. *Radiocarbon* 46: 1029–1058.

Rockman, M. 2003. Knowledge and learning in the archaeology of colonization. In: Rockman, M., Steele, J. (eds) *Colonization of unfamiliar landscapes: the archaeology of adaptation*, 3–24. Routledge, London.

Rosén, C. 1999. Fattigdomens arkeologi: reflektioner kring torpens och backstugornas arkeologi. In: Artelius, T., Englund, E., Ersgård, L. (eds) *Kring västsvenska hus: boendets organisation och symbolik i förhistorisk och historisk tid*, 99–107. Gotarc Serie C. Arkeologiska skrifter 22. Department of Archaeology, Gothenburg University.

Punt, W. (ed.) 1976-1996. *The Northwest European Pollen Flora 1-7.* Elsevier, Amsterdam.

Rubensson, L. 2000. Det småländska blästbruket och de arkeologiska spåren. In: Larsson, L.-O. & Rubensson, L. (eds) *Från blästbruk till bruksdöd: småländsk järnhantering under 1000 år.* II. Jernkontorets Bergshistoriska Skriftserie 35. Stockholm.

Ruddiman, W. F. 2005. *Plows, plagues, and petroleum: how humans took control of climate.* Princeton University Press, Princeton.

Sandnes, J. 1981. Settlement developments in the Late Middle Ages (approx. 1300–1540). In: Gissel, S., Jutikkala, E., Österberg, E., Sandnes, J., Teitson, B. (eds) *Desertion and land colonization in the Nordic countries c. 1300–1600: comparative report from the Scandinavian research project on deserted farms and villages,* 78–114. Almqvist & Wiksell, Stockholm.

Schoch, W. 1986. Wood and charcoal analysis. In: Berglund, B. E. (ed.) *Handbook of Holocene palaeoecology and palaeohydrology,* 619–626. Wiley, Chichester.

Schofield, J. E., Waller, M. P. 2005. A pollen analytical record for hemp retting from Dungeness Foreland, UK. *Journal of Archaeological Science* 32: 715–726.

Schweingruber, F. H. 1976. *Prähistorisches Holz: die Bedeutung von Holzfunden aus Mitteleuropa für die Lösung archäologischer und vegetationskundlicher Probleme.* Academica Helvetica 2. Paul Haupt, Bern.

Schweingruber, F. H. 1978. *Mikroskopische Holzanatomie: Formenspektern mitteleuropäischer Stamm- und Zwerghölzer zur Bestimmung von rezentem und subfossilem Material.* Paul Hapt, Bern.

Skansjö, S. 1997a. Örkelljunga–Fagerhult under medeltid och 1600-tal. In: Wallin, L., Olsson, M., Connelid, P., Karsten, P., Knarrström, B., Lagerås, P., Mattisson, A., Olsson, M., Skansjö, S. *Arkeologisk utredning från Örkelljunga till länsgränsen. Särskild arkeologisk utredning steg 1, väg E4, förbi Örkelljunga (Eket–Värsjö) och förbi Fagerhult (Värsjö–Köphult) samt väg 24, delen Bälinge–Västra Spång, Skåne,* 58–87. Riksantikvarieämbetet, UV Syd Rapport 1997:58.

Skansjö, S. 1997b. *Skånes historia.* Historiska Media, Lund.

Skog, G., Regnéll, J. 1995. Precision calendar-year dating of the elm decline in a Sphagnum-peat bog in southern Sweden. *Radiocarbon* 37: 197–202.

Sköld, E. 2006. *Kulturlandskapets förändringar inom röjningsröseområdet Yttra Berg, Halland: en pollenanalytisk undersökning av de senaste 5000 åren.* Examensarbeten i Geologi vid Lunds universitet, Kvartärgeologi 196. (Examination paper in Quaternary geology from Lund University.)

Söderberg, J., Myrdal, J. 2002. *The agrarian economy of sixteenth-century Sweden.* Acta Universitatis Stockholmiensis. Stockholm Studies in Economic History 35. Almqvist & Wiksell, Stockholm.

Stenström, J., Forshed, N. 2004. *Ljunghedar: historia, ekologi och arter.* Länsstyrelsen i Halland.

Subba Reddi, C., Reddi, N. S. 1986. Pollen production in some anemophilous angiosperms. *Grana* 25: 55–61.

Sugita, S. 1994. Pollen representation of vegetation in Quaternary sediments: theory and method in patchy vegetation. *Journal of Ecology* 82: 881–897.

Sugita, S., Gaillard, M.-J., Broström, A. 1999. Landscape openness and pollen records: a simulation approach. *The Holocene* 9: 409–421.

Sykes, M. T., Prentice, I. C. 1996. Climate change, tree species distributions and forest dynamics: a case study in the mixed conifers/northern hardwoods zone of northern Europe. *Climatic Change* 34: 161–177.

Tainter, J. A. 1988. *The collapse of complex societies.* Cambridge University Press, Cambridge.

Thompson, R., Oldfield, F. 1986. *Environmental magnetism.* Allen & Unwin, London.

Vestbö-Franzén, A. 1997. Aspekter på odling: jordbruk och odlingslandskap i Jönköpings län under förhistorisk tid och medeltid. In: Nodström, M., Varenius, L. (eds) *Det nära förflutna: om arkeologi i Jönköpings län,* 194–211. Småländska kulturbilder 1997. Jönköpings läns museum, Jönköping.

Vestbö-Franzén, A. 2004. *Råg och rön: om mat, människor och landskapsförändringar i norra Småland, ca 1550–1700.* Jönköpings läns museum, Jönköping.

Viklund, K. 2003. Att skilja agnarna från vetet: spår av forntida sädeshantering i södra Halland. *Utskrift* 7: 76–85.

Vuorela, I. 1973. Relative pollen rain around cultivated fields. *Acta Botanica Fennica* 102: 1–27.

Walden, J., Oldfield, F., Smith J. 1999. *Environmental magnetism: a practical guide.* Technical Guide 6. Quaternary Research Association, London.

Wallin, J.-E. 2004. Människan och landskapet i Halland: en miljöhistorisk studie över brons- och järnåldersbygd, baserad på pollenanalyser. In: Carlie, L., Ryberg, E., Streiffert, J., Wranning, P. (eds) *Landskap i förändring–hållplatser i det förgångna:*

artiklar med avstamp i de arkeologiska undersökningarna för Västkustbanans dubbelspår förbi Falkenberg i Halland, 43–54. Halland Länsmuseer, Halmstad/Riksantikvarieämbetet, Stockholm.

Webb, T. III 1986. Is vegetation in equilibrium with climate? How to interpret late-Quaternary pollen data. *Vegetatio (Plant Ecology)* 67: 75–91.

Wedberg, V. 1981. Sven Nöjds studier i tidig nordskånsk järnhantering med inledande beskrivning över det skånsk-halländska järnområdet. Institute of Archaeology, Stockholm University. (Unpublished undergraduate seminar paper.)

Welinder, S. 1975. *Prehistoric agriculture in eastern Middle Sweden*. Acta Archaeologica Lundensia, Series in 8° Minore 4. Gleerup, Lund.

Welinder, S. 1992. *Människor och landskap*. Societas Archaeologica Upsaliensis. Aun 15. Department of Archaeology, Uppsala University.

Wendland, W. M., Bryson, R. A. 1974. Dating climatic episodes of the Holocene. *Quaternary Research* 4: 9–24.

Widgren, M. 1997. *Fossila landskap: en forskningsöversikt över odlingslandskapets utveckling från yngre bronsålder till tidig medeltid*. Kulturgeografiskt seminarium 1/97. Kulturgeografiska institutionen, Stockholms universitet.

APPENDIX 1: METHODS

·········

Pollen analysis

Peat and gyttja sequences were sampled for pollen analysis with what is known as a Russian corer, with a diameter of 10 cm (Jowsey 1966, Aaby and Digerfeldt 1986). All stratigraphies were documented in the field and all cores were brought to the laboratory as 1 m segments and stored in a cold room. Subsampling for pollen analysis, radiocarbon dating, etc., took place indoors at the laboratory.

Pollen preparation followed standard methods (e.g. Erdtman 1960, Berglund and Ralska-Jasiewiczowa 1986, Moore et al. 1991). It included the addition of *Lycopodium* tablets to volume-specific samples, sieving (mesh: 250 µm), treatment with 10% NaOH, 40% HF, 10% HCl, acetolysis with 1 part H_2SO_4 to 9 parts $C_4H_6O_3$, and final mounting in glycerine. Pollen was counted at a magnification of 400–500 (standard) and 1000–1250/phase contrast (critical identifications). Approximately 1000 pollen grains were counted in each sample. The number of microscopic charcoal particles with a maximum diameter >25 µm found in the slides during the pollen counting was also registered.

Pollen identification was aided by pollen reference collections at the Department of Geology department, Quaternary Sciences, Lund university, and the Swedish National Heritage Board, Archaeological Excavations Department in Lund, as well as by identification keys and other relevant literature (e.g. Punt 1976–1996, Fægri and Iversen 1979, Moore et al. 1991, Reille 1995, 1998, 1999).

Special attention was paid to the complicated but important identification and distinction of different cereal pollen types. Size criteria used for the identification (i.e. grain size, annulus diameter, and spore diameter) was based on the measurements and results presented by Andersen (1979). As recommended by Fægri and Iversen (1979; see also Lagerås 1996b), these measurements were multiplied by 1.2 to fit pollen mounted in glycerine (Andersen used silicone oil for mounting). The identification of cereal pollen was, however, not based solely on measurements and shape, but also on sculpture characteristics examined using phase contrast (Beug 1961).

The pollen analysis was performed by Leif Björkman at Lund University and by Nils-Olof Svensson and myself at the National Heritage Board.

The total pollen diagrams, as well as some of the special diagrams, were prepared using the computer programs TILIA 2 and TGView 2.0.2 (E. C. Grimm, Illinois State Museum, U.S.A.), while some of the special diagrams were prepared using Grapher 4.0. The total diagrams were constructed mainly in accordance with recommendations by Berglund and Ralska-Jasiewiczowa (1986). In the left part of the diagrams, next to the time and depth scales, there is a cumulative area diagram showing the proportions between the following four plant groups: trees, shrubs (includes some small trees), dwarf shrubs (mainly Ericaceae), and herbs. After that, all identified pollen taxa are presented with individual graphs, which are ordered in the same groups as in the cumulative area diagram.

Within the tree group, the individual taxa are presented in their approximate Holocene immigration order, i.e. according to their first appearance during the last 11000 years approximately. Within other groups (shrubs, dwarf shrubs, and herbs), taxa are presented more or less in systematic order according to the most influential Swedish flora (Krok and Almquist 1994). An exception is that all cultivated taxa are presented together in the left part of the herb group.

Percentages of terrestrial pollen as well as non-terrestrial pollen and spores are based on the sum of terrestrial pollen in each sample. In the diagram from Grisavad, *Alnus* and *Myrica* are excluded from the pollen sum because of their strong local over-representation.

It is important to note that all pollen diagrams in the book are plotted on a linear time scale rather than a linear depth scale. As peat accumulation rates always vary through time, it is not possible to have both a linear time scale and a linear depth scale in the same diagram. I have chosen to use linear time scales because they increase the readability and also the possibility to compare different diagrams with each other. This approach has been possible due to the relatively large number of radiocarbon-dated levels, and the careful selection of dating material. The time scales show calibrated years A.D. To facilitate comparison, all the diagrams cover the same time interval, i.e. the last 2000 years.

Mineral magnetic analysis

The pollen-analysed cores were also subject to mineral magnetic analysis in order to trace soil erosion (see Thompson & Oldfield 1986 or Walden et al. 1999 for an overview of mineral magnetic techniques; see also Lagerås & Sandgren 1994). The mineral magnetic parameter that was measured was SIRM (Saturated Isothermal Remanent Magnetization). It is a measure of the ferromagnetic concentration in a sample, which reflects the concentration of mineral particles.

Peat and gyttja sequences were subsampled contiguously into polystyrene boxes at 2 cm intervals. Samples were magnetized in a magnetic field of 1 Tesla obtained by a Redcliff Pulse Magnetizer, after which the magnetic remanence was measured in a Molspin Spinner Magnetometer. Samples were dried and weighed to enable calculation of mass-specific parameters.

The mineral-magnetic analysis was performed by Per Sandgren at Lund University.

Radiocarbon dating

Reliable and independent chronologies are fundamental for the interpretation of pollen diagrams and in particular when they are to be compared with other data, for instance from archaeology or written records. For each of the pollen diagrams presented in this book a detailed and independent absolute chronology was established based on radiocarbon dates.

Most important in radiocarbon dating is the material used for the analysis (e.g. Olsson 1986, 1991). In earlier studies it was common to use bulk samples, which in many cases gave erroneous results. In gyttja sequences bulk samples usually gave dates that were too old, due to the so-called reservoir effect, while in peat they sometimes gave dates that were too recent, due to down-growing roots from younger layers.

In the present study, depending on the character of the deposits and the degree of humification, different types of material have been sampled for radiocarbon dating. In the gyttja sequence from Östra Ringarp the material used for dating was terrestrial plant macrofossils (seeds, catkins scales, etc.), while in the peat sequences most of the dating was performed on peat samples that had been carefully cleared from roots and wood (when possible *Sphagnum* moss plants were used). In some cases pollen concentrates were also used (cf. Regnéll 1992, Regnéll & Everitt 1996).

All samples were dated using the AMS (Accelerator Mass Spectrometer) technique. The dating was performed by the Ångström Laboratory at Uppsala University, the Radiocarbon Dating Laboratory at Lund University, and the Poznan Radiocarbon Laboratory.

Radiocarbon dates were calibrated to calendar years using the computer program OxCal 3.10 and atmospheric data from Reimer et al. (2004). Based on the probability distribution for each calibrated date, together with interpretation of the lithology and expected accumulation rates, time/depth graphs were constructed using linear interpolation. These time/depth graphs were then used to construct the time-scales of the pollen diagrams.

Macroscopic charcoal analysis

Macroscopic charcoal pieces were collected from different archaeological features during the excavations. These features were pits, hearths, post-holes, clearance cairns, charcoal production sites, iron furnaces and slag heaps, tar pits, etc. Most charcoal samples were collected from cleaned midsections, and their stratigraphical positions within the archaeological features were carefully documented.

Altogether 140 samples were subject to charcoal analysis (i.e. identification to tree species or genera). Most samples included several pieces and the total number of identified charcoal pieces was approx. 3300. Of these, 141 pieces were radiocarbon-dated (see table 13 in *Appendix 2*).

The charcoal analysis followed standard procedures (Schweingruber 1976, 1978, Bartholin et al. 1981, Schoch 1986, Bartholin and Berglund 1992), using a microscope with reflected light and a magnification of ×30–300. A reference collection of recent wood and charcoal was used. Most of the identified charcoal pieces were only a few millimetres in size.

The charcoal analysis was performed by Thomas Bartholin, National Museum of Denmark.

APPENDIX 2: RESULTS AND INTERPRETATIONS SITE BY SITE

· · · · · · · ·

Overview of data from different sites

In an early phase of the project extensive reconnaissance coring was carried out in numerous peatlands and lakes all along the exploitation area for the new motorway (Lagerås 2003a). Coring was carried out at more than one hundred coring points, stratigraphies were documented, and key layers were radiocarbon-dated (primary data from the reconnaissance coring were published in two reports in Swedish; Lagerås 1997, 2003b). Based on the results – together with a valuation of the distribution, location, and character of documented archaeological features – four sites were selected for detailed pollen analyses: Östra Ringarp, Grisavad, Värsjö Utmark, and Bjärabygget (figure 2). The main pollen diagrams from these detailed studies are presented site by site below.

In addition to pollen analysis, cores from the same sites were also subject to mineral-magnetic analyses in order to trace periods of soil erosion. It turned out, however, that only the gyttja sequence from Östra Ringarp gave useful results, while the peat sequences from the other three sites showed a magnetic signal that was too weak. The result from Östra Ringarp is presented in connection to the pollen diagram below.

Another important source material, which contributes to the interpretations and discussions, is macroscopic charcoal. It was collected from archaeological features – such as hearths, clearance cairns, bloomery furnaces, charcoal production sites, etc. – at a large number of excavation sites along the investigation area. Hence, the charcoal data (radiocarbon dates and tree-species identification) are not restricted to the four pollen-analysed sites, and they are presented in a separate section below.

Östra Ringarp

The sampling site Östra Ringarp is situated in the southwestern part of the investigation area, two kilometres southeast of the city of Örkelljunga and less than one kilometre southeast of Lake Hjälmsjön. The local area is characterized by a gently hummocky landscape of sandy till and glaciofluvial sand and gravel, at an altitude of about 100 m above sea level. The landscape is to a large degree forested, but there are also pastures and some arable land in the vicinity of the sampling site.

The sampled peatland is a small fen with a dense tree vegetation of birch (*Betula* sp.), pine (*Pinus syvestris*) and spruce (*Picea abies*), and ground vegetation with sedges (*Carex* sp.) and hare's-tail cottongrass (*Eriophorum vaginatum*). Coring revealed that the fen is a former lake, which was finally choked-up and overgrown only about two hundred years ago. The approximate diameter of the lake was 150 m. Based on studies on pollen representation (Sugita 1994), the relevant pollen source area of a pollen diagram from a lake of this size can be expected to have a radius of approx. 300–500 m. Hence, the diagram from this site would mainly reflect vegetation and land-use within a few hundred metres, which is a suitable scale for this type of study.

Very close on the western side of the peatland a farm is situated. It has some small arable fields but mainly pastures grazed by sheep. According to the oldest detailed map of the farm, which is dated to 1818–1824, the buildings at that time were situated at the same place as today (figure 10). It is interesting to note the shape of the arable fields on the old map. They are to a large degree arranged in a strip field pattern, with narrow elongated fields running in a west-easterly direction, most clearly visible in the area northwest of the farm. This field pattern has been noted on old maps of several farms in the investigation area, and also in other parts of southern Sweden, and it has been interpreted as originating in the Viking Period or in the Middle Ages (e.g. Connelid 2002).

Except for some test-trenching, no archaeological excavation was carried out at Östra Ringarp within the project. But there has been an older excavation on the site. In 1953 a furnace and some other remains of bloomery iron production were excavated in close connection to the farm and the peatland (Wedberg 1981, Englund 1995). Charcoal from the furnace was radiocarbon-dated to the 13th and 14th centuries, i.e. the Middle Ages, which gives support to a medieval origin of the farm.

At the coring point for pollen analysis there was a total thickness of organogenic deposits of more than six metres. The stratigraphy was dominated by gyttja deposited in the former lake, and only the upper 57 cm was peat (table 5). The transition from gyttja to peat was dated to c. A.D. 1800. Based on a reconnaissance analysis (Björkman & Ekström 2002), the upper 360 cm of the sequence was subjected to a detailed pollen analysis. In this analysis the elm decline was identified at 315 cm and it could be concluded that the analysed sequence of 360 cm covered the last approximate 7000 years. In order to get a good time-resolution and readability of the time period that is in focus for this book, only the upper part of the diagram which represents the last 2000 years is presented here (figure 49; however, see also figure 5). In connection to the pollen analysis ten levels were radiocarbon-dated (table 6).

With the aim to identify anthropogenic soil erosion, the pollen-analysed core was subject to mineral magnetic analysis. The measured parameter was SIRM (Saturated Isothermal Remanent Magnetization), and the result is presented in figure 48 and in connection to the pollen diagram in figure 49. The peat sequence in the upper part of the core (0–60 cm) was not analysed, as to little material was left after the subsampling for pollen analysis and radiocarbon dating.

Table 5. Peat and gyttja stratigraphy of the pollen-analysed sequence at Östra Ringarp.

Depth (cm)	Description
0–20	Sphagnum peat, weakly to moderately humified
20–57	Carex peat, moderately humified
57–75	Coarse detritus gyttja
75–516	Fine detritus gyttja
516–600	Clay gyttja

Table 6. *Radiocarbon dates from the pollen-analysed sequence at Östra Ringarp.*

Depth (cm)	Lab.no	C14 BP	Cal 2s (max–min)	Dating material
21.5–25	LuS-5928	125 ± 50	A.D. 1670–1960	Sphagnum peat cleared from roots and wood
45.5–51	LuS-5929	110 ± 50	A.D. 1670–1960	Carex peat cleared from roots and wood
51–52	Ua-26416	65 ± 35	A.D. 1680–1960	Carex peat cleared from roots and wood
57–59	LuS-6217	190 ± 50	A.D. 1640–1960	Terrestrial plant macrofossils
65–70	Ua-26417	490 ± 40	A.D. 1320–1480	Terrestrial plant macrofossils
75–80	LuS-6216	580 ± 50	A.D. 1290–1430	Terrestrial plant macrofossils
100–105	LuS-6215	200 ± 50	A.D. 1630–1960	Terrestrial plant macrofossils
115–120	LuS-6214	1015 ± 50	A.D. 890–1160	Terrestrial plant macrofossils
135–140	LuS-6213	1460 ± 50	A.D. 430–670	Terrestrial plant macrofossils
218–222	LuS-6211	3560 ± 60	2120–1740 B.C.	Terrestrial plant macrofossils

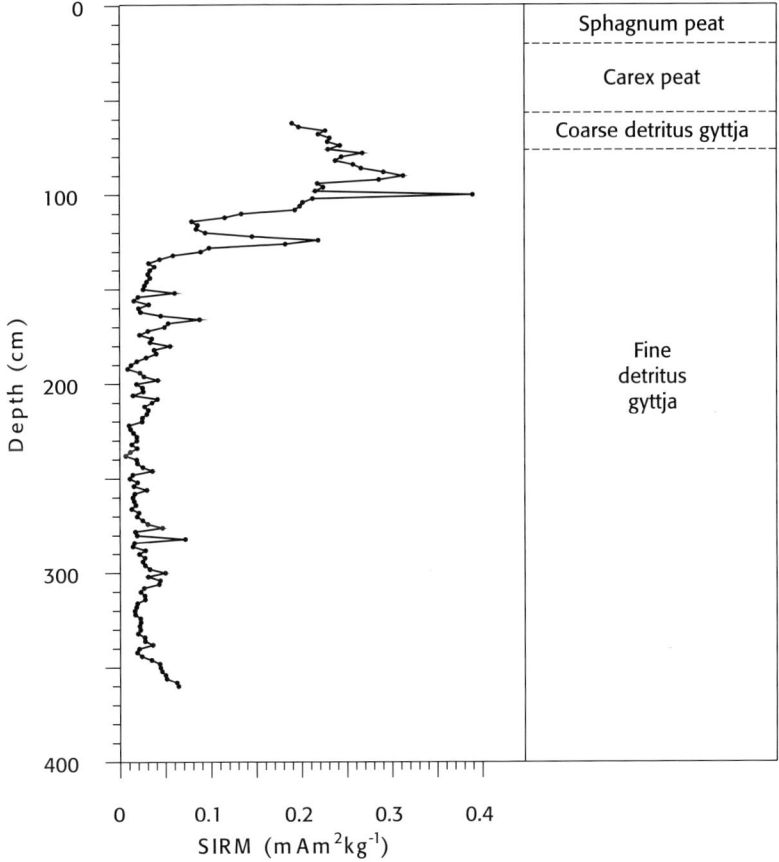

Fig. 48. *Weight-specific SIRM values from Östra Ringarp plotted on a linear depth scale.*

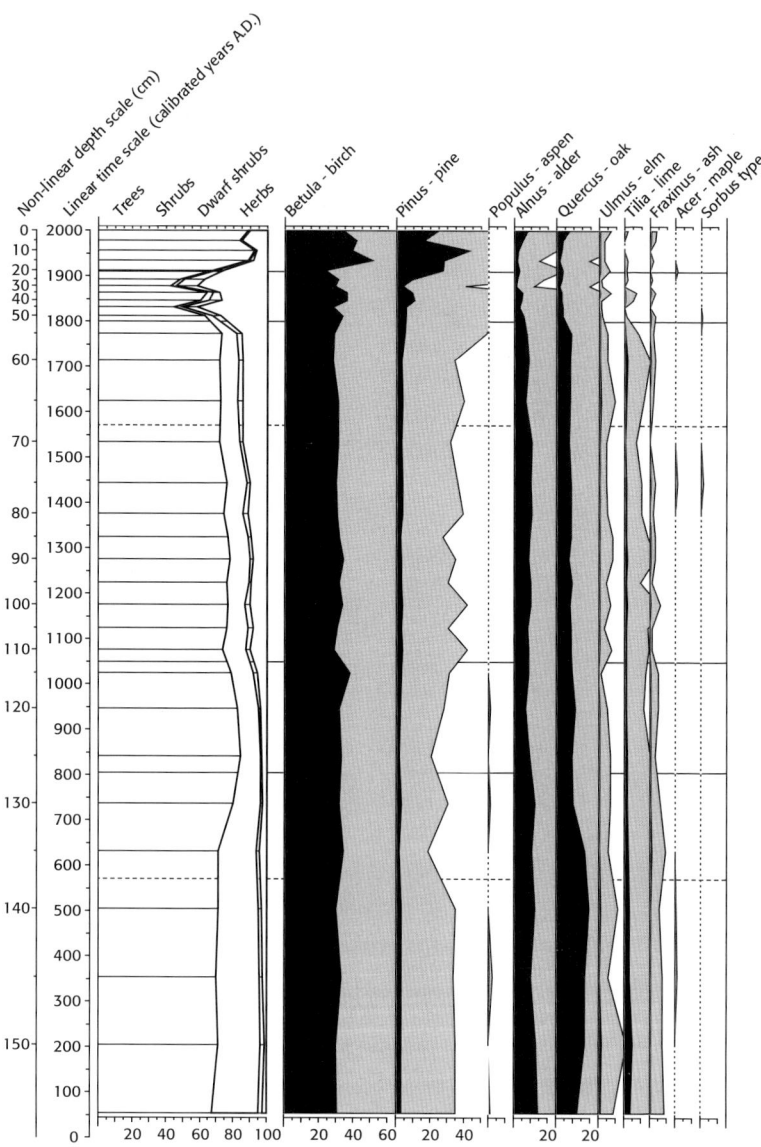

Fig. 49. Total pollen diagram from Östra Ringarp (56°16'N, 13°19E) with all identified pollen taxa. The graphs show pollen percentages (black) and ten times exaggeration (shaded), and they are based on the total pollen sum. Taxa presented to the left of the pollen sum in the diagram are included in the sum. Analysed by Leif Björkman.

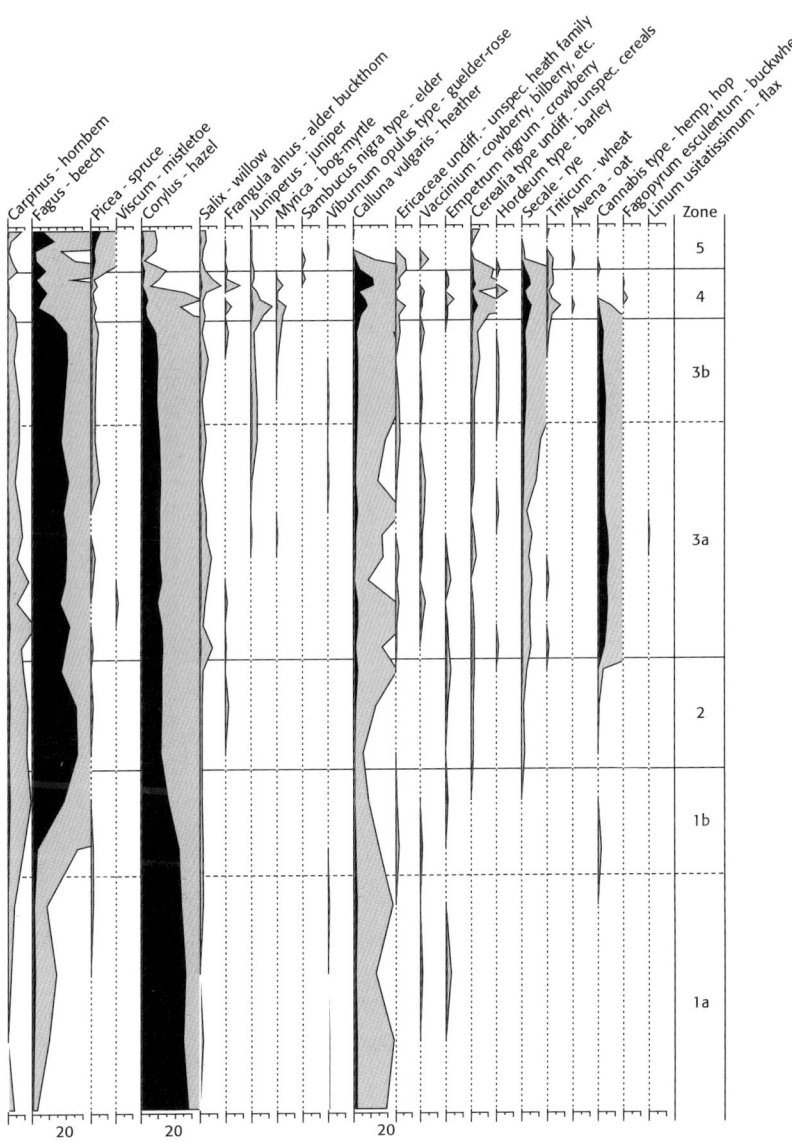

Fig. 49. Continued.

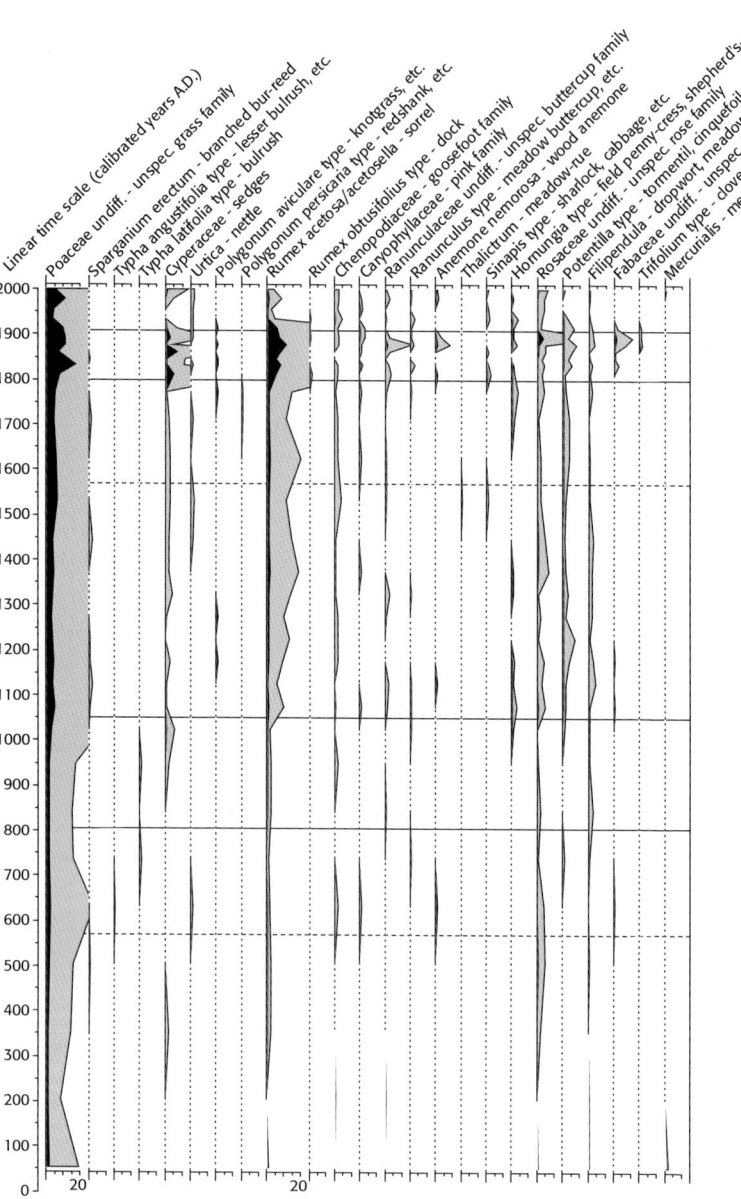

Fig. 49. Continued.

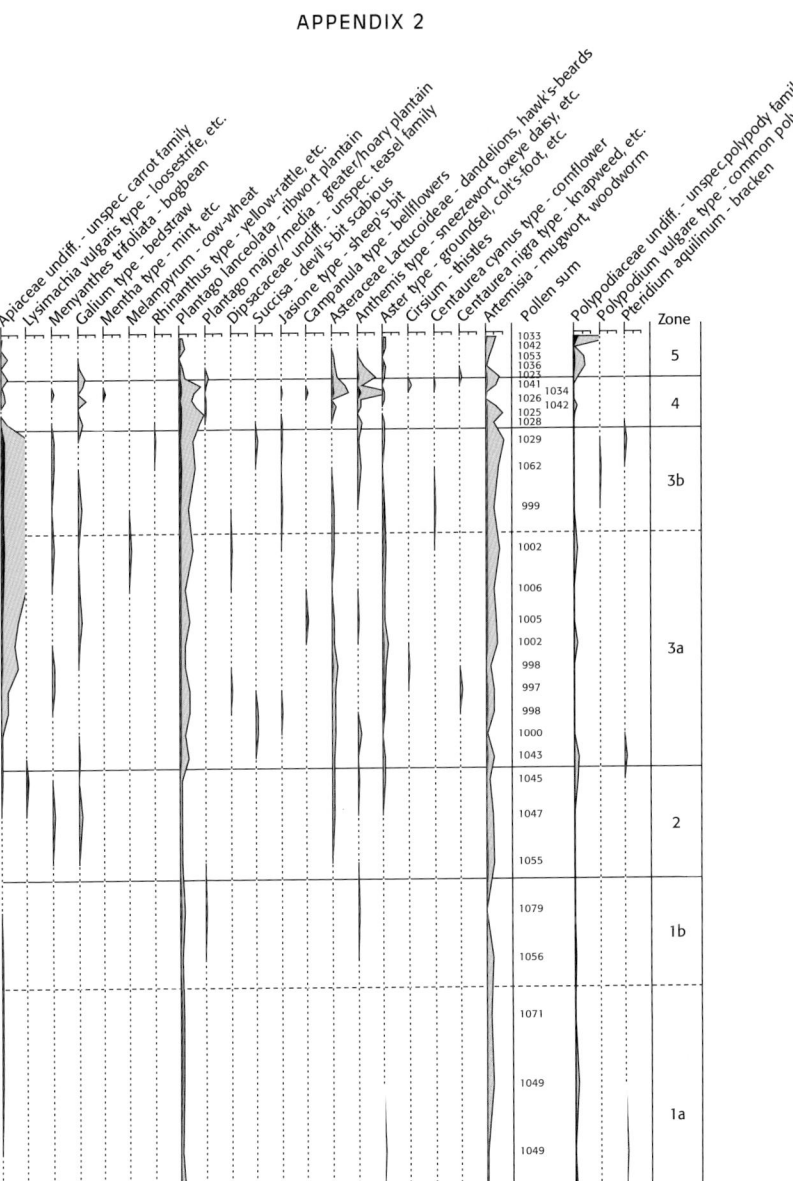

Fig. 49. Continued.

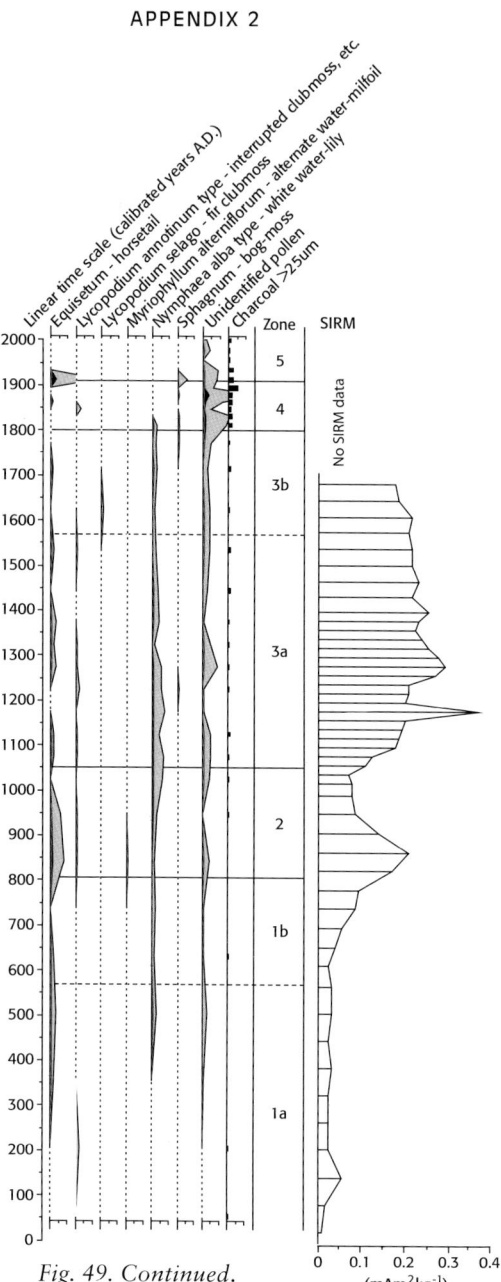

Fig. 49. Continued.

The pollen diagram has been divided into five zones which here will be briefly described. These zones are pollen assemblage zones (Hedberg 1976, Berglund & Ralska-Jasiewiczowa 1986), but with emphasis on human impact indicators. As the diagram is plotted on a linear time scale the zones are sometimes in the text also referred to as periods.

Zone 1 – Wood pasturage (>0 B.C.–c. A.D. 800): Zone 1 reflects a forested landscape. Dominating tree taxa on dry ground were *Quercus* (oak) and *Tilia* (lime), and the more light-demanding *Corylus* (hazel) and *Betula* (birch). *Alnus* (alder) and probably *Betula* thrived on wet soils and may have bordered the sampled lake. In the upper part of the zone (distinguished as subzone 1b) there is a strong increase in *Fagus* (beech), which marks the local establishment and expansion of a beech forest. The expansion lasted until A.D. 800 when *Fagus* reached its maximum values. The increase in *Fagus* is matched by a decrease in *Quercus* (oak), *Tilia* (lime), and *Corylus* (hazel) – taxa that obviously lost in competition with the dominant and shade-tolerant *Fagus*.

Throughout zone 1 there is a continuous curve for *Plantago lanceolata* (ribwort plantain), which together with a regular occurrence of other herb taxa indicate grazing. The high frequencies of tree pollen and the low frequencies of Poaceae undiff. (grass) indicate, however, that grazing only resulted in minor openings in the forest. The land-use is interpreted as wood pasturage, i.e. extensive grazing and browsing in a mainly forested landscape. [A more or less continuous occurrence of *P. lanceolata* begins at a level dated to c. 1700 B.C. (see figure 5), which indicates that wood pasturage started as early as the Early Bronze Age.]

Zone 2 – Temporary cultivation in beech forest (c. A.D. 800–1050): As in the previous zone, there are indications of wood pasturage throughout zone 2. But even though the landscape was mainly forested there is also some cereal pollen from this zone which indicates cultivation. These single pollen grains of

Secale (rye) and Cerealia type undiff. (unspec. cereals) probably originate from temporary cultivation in clearings in the forest. In addition, the SIRM values are relatively high, which probably reflect increased soil erosion due to the clearings.

Zone 3 – Permanent cultivation by local farm (c. A.D. 1050–1800): The beginning of zone 3 marks a major change in vegetation and the birth of a relatively open cultural landscape, which indicates the local establishment of a permanent farm. Relatively high percentages of Poaceae undiff. (grass) and a large number of herb taxa reflect pastures and probably hay meadows. Cultivation is reflected in the occurrence of Cerealia type undiff. (unspec. cereals), *Secale* (rye), *Hordeum* (barley), *Triticum* (wheat), *Linum usitatissimum* (flax), and *Cannabis* type (hemp). The last mentioned pollen type may be either *Cannabis sativa* (hemp) or *Humulus lupulus* (hops), but the high frequencies show that it must be hemp which possibly has been retted in the lake.

The SIRM values are relatively high, which probably reflect soil erosion in arable fields.

The overall impression from the pollen diagram is a continuous land-use with very little change throughout zone 3, i.e. a period of approx. 700 years. However, at a level dated to c. A.D. 1550, there is an increase in *Secale*, *Juniperus* (juniper), Poaceae undiff., Apiaceae undiff. (unspec. carrot family) and some other taxa, which may indicate a slight intensification of both cultivation and grazing. This upper part of zone 3 is distinguished as subzone 3b.

Zone 4 – Permanent cultivation and heathland (c. A.D. 1800–1900): After a long period of continuity, the beginning of zone 4 reflects some major changes in land-use. The cultivation of *Cannabis* (hemp) stopped while there was an increase in the cultivation of *Secale* (rye) and *Triticum* (wheat), and also an introduction of *Avena* (oats) and *Fagopyrum esculentum* (buckwheat). Increasing frequencies of Poaceae undiff. (grass) indicate an expansion of pastures and probably hay meadows, and the relatively high values for *Calluna* (heather) show that some of the pastures developed into heathland. Furthermore, the occurrence in this zone of pollen grains of Fabaceae undiff. (unspec. bean family) and *Trifolium* type (clover) may be connected to crop rotation with nitrogen-fixing Fabaceae.

Zone 5 – Pasture and silviculture (c. A.D. 1900–present): Zone 5 is characterized by a decrease in *Calluna* (heather) and cultivation indicators, and an increase in *Pinus* (pine), *Betula* (birch), and *Picea* (spruce). Obviously, the extent of arable fields decreased and *Calluna* heathland was planted with forest. However, the zone shows high frequencies of Poaceae undiff. (grass), which fits in well with the present situation – the farm that still exists on the site is based on sheep husbandry and there are grazed pastures in close connection to the peatland.

Grisavad

The sampling site Grisavad is situated in the southwestern part of the investigation area, in a gently hummocky landscape of glaciofluvial sand and gravel, at an altitude of about 100 m above sea level. A little less than one kilometre to the west is Lake Hjälmsjön and in the north runs the small River Lärkesholmsån. Grisavad is situated 1.6 km northeast of the site Östra Ringarp presented above. The local area is forested, mainly with planted spruce (*Picea abies*) and beech (*Fagus sylvatica*), but open agricultural land is found within two kilometres, both to the south at Östra Ringarp and to the north at Östra Spång.

The sampled peatland is a bog with tree vegetation of pine (*Pinus sylvestris*) and birch (*Betula* sp.) and ground vegetation dominated by heather (*Calluna vulgaris*) and hare's-tail cottongrass (*Eriophorum vaginatum*). The coring point for pollen analysis was situated 20 metres from the edge, facing the remains of a croft which was archaeologically investigated within the project (figure 12). Visible features were the remains of a small dwelling house, outhouses, a well and some stone walls. The buildings had probably been timber-framed with a thatched or turfed roof, but now only the stones of wall foundations and a collapsed chimney together with some cellar pits remain. In connection with the croft there were also field terraces, clearance cairns and ancient pathways. The remains were carefully mapped and a few square metres were excavated (A. Knarrström 2003).

A map from 1818 shows that the site at that time was situated on the outland between the settlements of Östra Ringarp in the south and Östra Spång in the north. There are no older preserved maps than the one from 1818 which can be used to trace Grisavad back in time, but the place-name Grisavad is mentioned earlier in the Swedish Civil Survey from 1671 (Sw: *jordrevningsprotokoll*) (Skansjö in prep.). Archaeological finds from the small-scale excavation were pottery, china, glass, nails, a comb, locks, coins, buttons, bones, etc. Most of these finds could be dated to the 19th century, but some of them could also be from the 18th or 17th centuries. By and large the archaeological finds thus confirmed the preliminary dating based on historical sources.

At the coring point there was a total peat thickness of 145 cm, with fen peat in the lower part and *Sphagnum* peat in the upper part (Table 7). Peat started to accumulate due to paludification a little more than two thousand years ago, according to the lowermost radiocarbon date. Unusually enough, the peat sequence was not compacted by drainage and showed a good temporal resolution all the way up to the surface. Based on a reconnaissance analysis of the upper 100 cm and three radiocarbon dates (Björkman & Ekström 2002), it was decided to

Table 7. Peat stratigraphy of the pollen-analysed sequence at Grisavad.

Depth (cm)	Description
0–35	Sphagnum peat, weakly humified
35–45	Sphagnum peat, moderately humified
45–56	Sphagnum peat, strongly humified
56–143	Fen peat
143–145	Fen peat/gyttja

Table 8. Radiocarbon dates from the pollen-analysed peat sequence at Grisavad.

Depth (cm)	Lab.no	C14 BP	Cal 2s (max–min)	Dating material
10.5–12	LuS-5930	65 ± 50	A.D. 1670–1960	Sphagnum moss
20–21	Ua-26398	175 ± 35	A.D. 1650–1960	Sphagnum peat cleared from roots and wood
36–37	LuS-5931	215 ± 50	A.D. 1520–1960	Sphagnum peat cleared from roots and wood
50–51	Ua-26399	705 ± 40	A.D. 1240–1400	Sphagnum peat cleared from roots and wood
58–59	LuS-5968	1120 ± 70	A.D. 710–1030	Pollen concentrate
99–101	Ua-26400	2070 ± 40	200 B.C.– A.D. 30	Fen peat cleared from roots and wood

use the entire sequence of 0–145 cm for a detailed pollen-analytical study (Björkman 2004). In connection to this final analysis six levels were radiocarbon-dated (see table 8).

The original pollen analysis covered the last approximate 3200 years, while the diagram presented here only covers the last 2000 years (figure 50). The diagram has been divided into seven zones which here will be briefly described. These zones are pollen assemblage zones (Hedberg 1976, Berglund & Ralska-Jasiewiczowa 1986), but with emphasis on human impact indicators. As the diagram is plotted on a linear time scale the zones are sometimes also referred to in the text as periods.

Zone 1 – Wood pasturage (>0 B.C.–c. A.D. 750): This lowermost zone in the diagram reflects a forested landscape, with in particular *Betula* (birch) but also some *Alnus* (alder) growing on the sampled peatland. Dominating taxa on dry ground were *Quercus* (oak), *Tilia* (lime) and *Corylus* (hazel). *Fagus* (beech), however, started to increase strongly in c. A.D. 300 and became the dominating taxa by the end of the period. The increase in *Fagus* in A.D. 300 coincides relatively well with an increase in Poaceae undiff. (grass), which may indicate that small-scale vegetation clearances at this time favoured the local establishment of *Fagus*. (The period that followed the increase in *Fagus* and Poaceae undiff. in c. A.D. 300 is distinguished as subzone 1b in the diagram.) Such interpretation is in line with the one put forward by Björkman (1996a), who argued that the establishment of *Fagus* woodlands in southern Sweden in some places was favoured by anthropogenic disturbances.

221

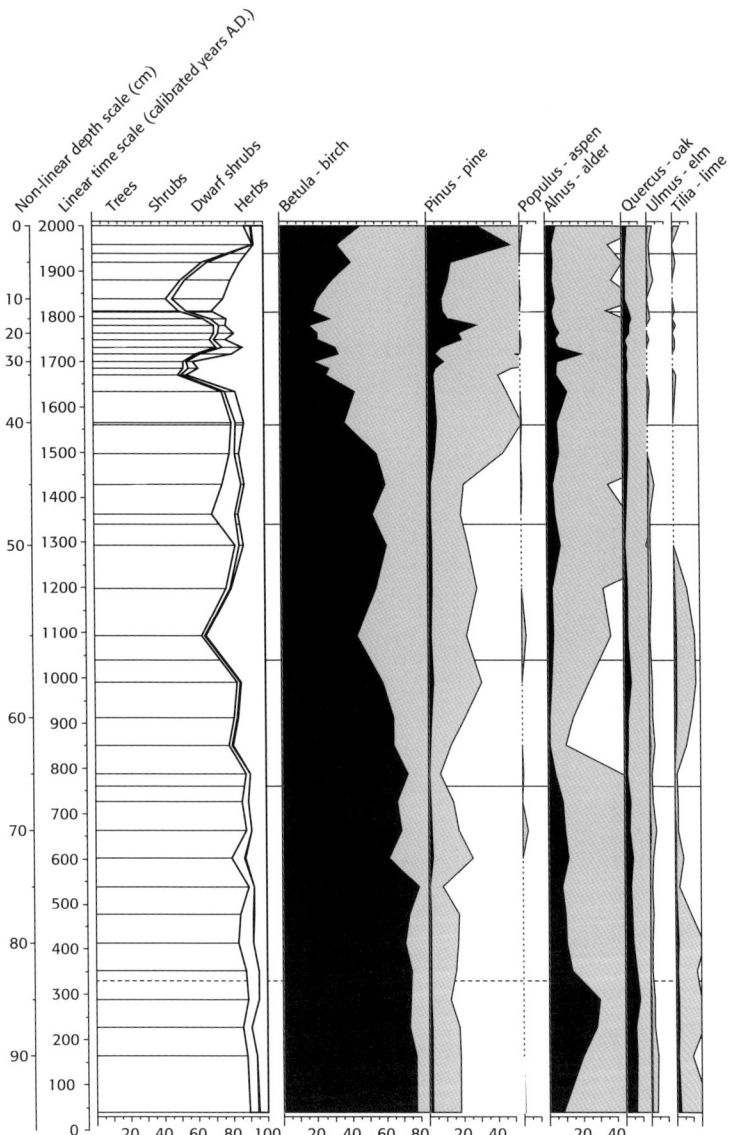

Fig. 50. Total pollen diagram from Grisavad (56°17'N, 13°20E) with all identified pollen taxa. The graphs show pollen percentages (black) and ten times exaggeration (shaded), and they are based on the total pollen sum. Taxa presented to the left of the pollen sum in the diagram are included in the sum. Analysed by Leif Björkman.

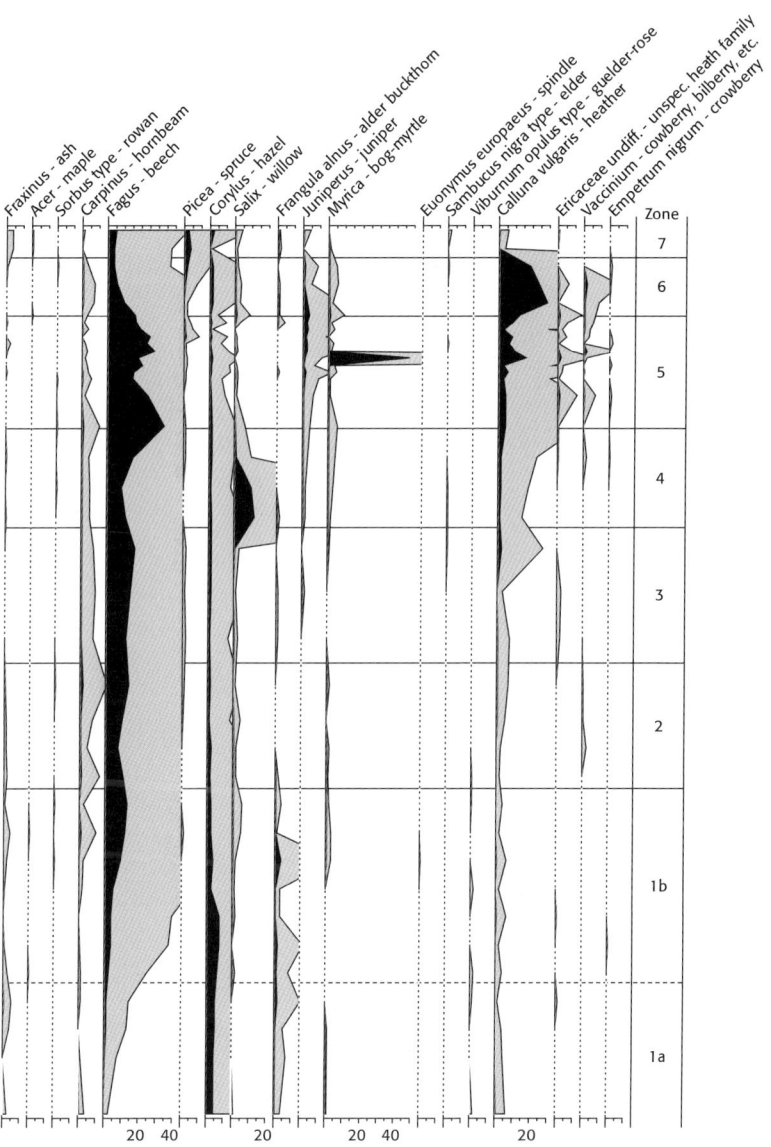

Fig. 50. Continued.

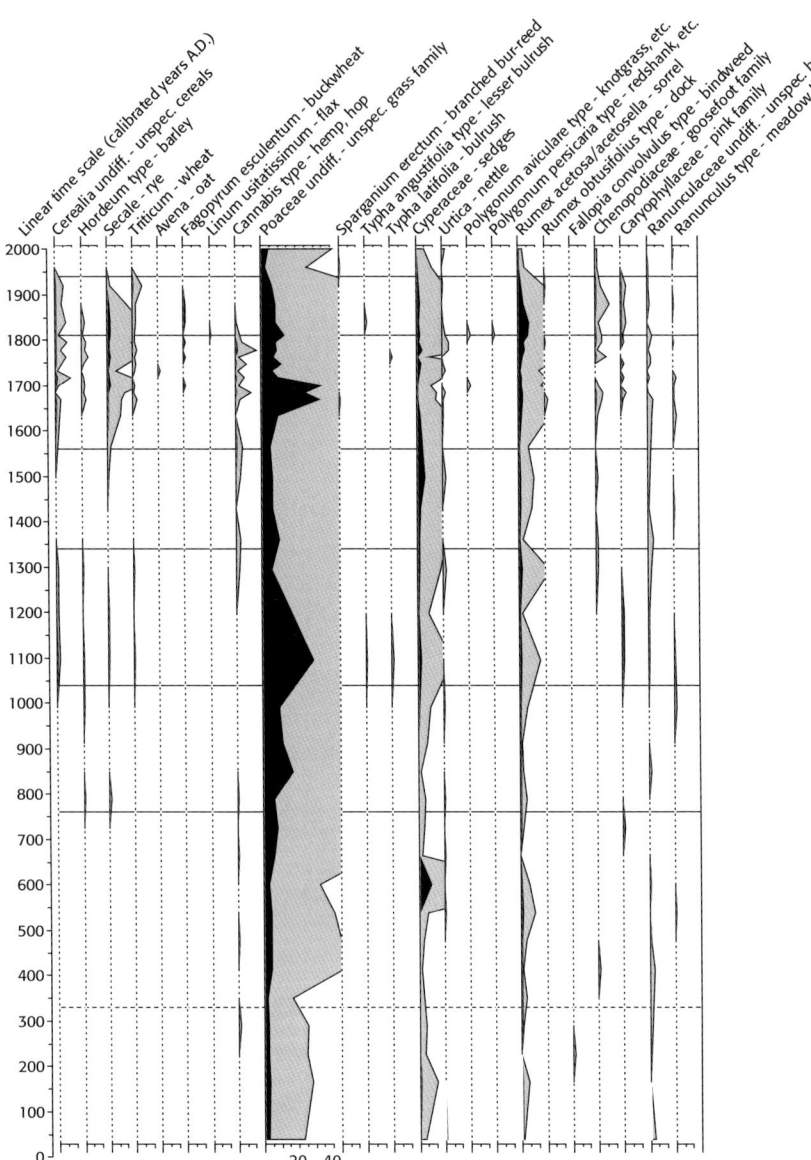

Fig. 50. Continued.

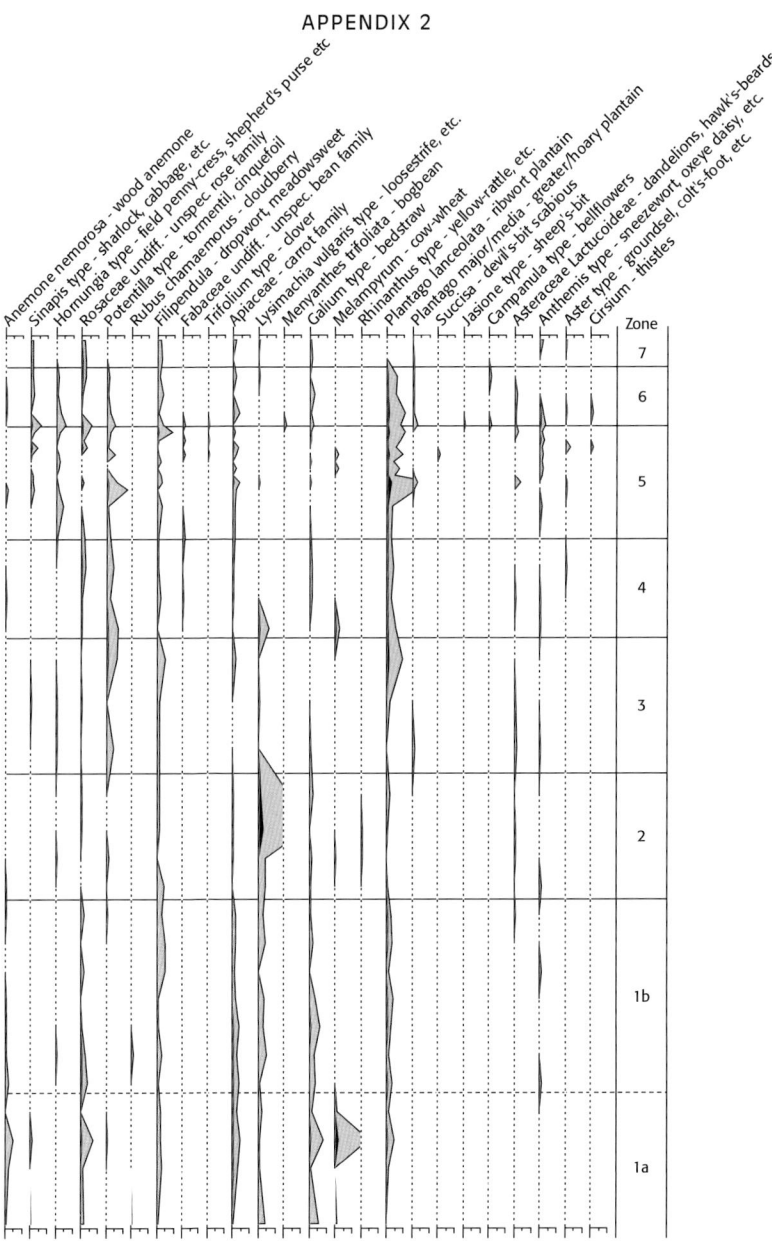

Fig. 50. Continued.

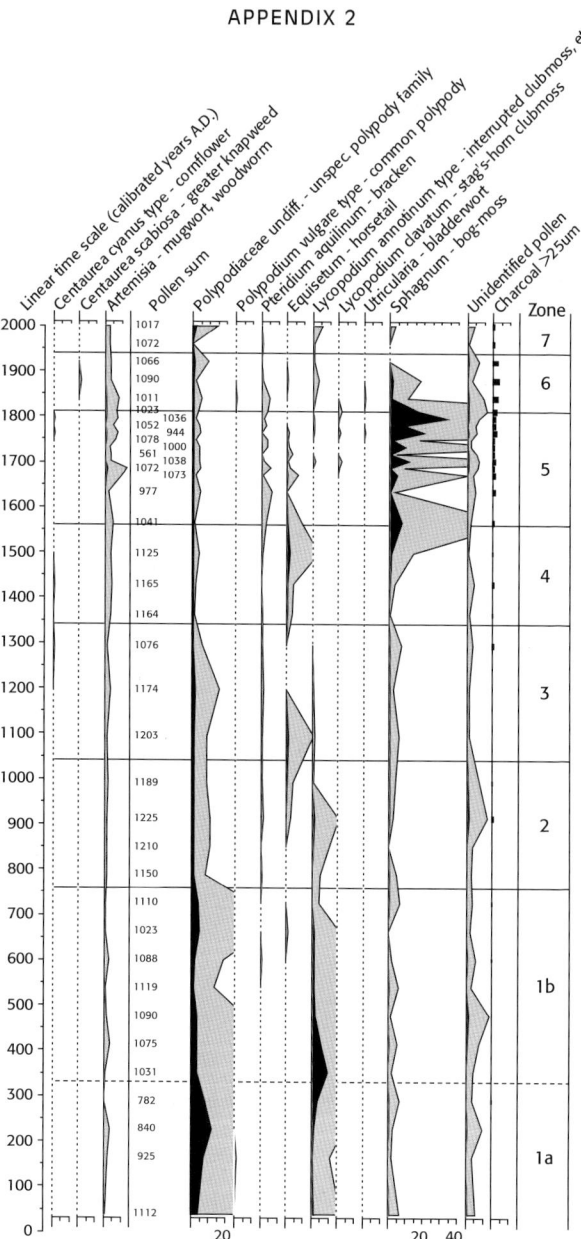

Fig. 50. Continued.

Throughout zone 1 there is a regular occurrence of herb taxa, for example a more or less continuous curve for *Plantago lanceolata* (ribwort plantain), which indicates grazing. The high frequencies of tree pollen and the low frequencies of Poaceae undiff. indicate, however, that grazing only resulted in minor openings in the forest.

Zone 2 – Temporary cultivation and grazing (c. A.D. 750–1050): The beginning of zone 2 is characterized by an increase in Poaceae undiff. (grass) and a first appearance of *Hordeum* (barley) and *Secale* (rye), which, together with a decrease in *Fagus* (beech) and *Quercus* (oak), reflects clearings and deforestation on dry ground to create fields and pastures. The decrease in *Alnus* (alder) indicates that tree vegetation was also cleared on wetlands. Obviously these changes reflected in the beginning of zone 2 reflect an agrarian expansion in the area. They may reflect the local establishment of a farm, but the cereal record is very sparse and discontinuous. More likely the area was used for grazing and temporary cultivation by one or several farms situated at some distance.

Zone 3 – Permanent cultivation by local farm (c. A.D. 1050–1350): Zone 3 is similar to zone 2 but has a more continuous cereal record and also shows higher values for *Rumex acetosa/acetosella* (sorrel) and some other weeds. At this time there was certainly a permanent farm in the local area, probably established by the beginning of the period, i.e. c. A.D. 1050. The farm cultivated *Hordeum* (barley), *Secale* (rye) and *Triticum* (wheat), and it had open pastures with field vegetation dominated by grasses. Some of these grasslands may also have been used for hay mowing. In the later part of the period there is a decrease in Poaceae undiff. (grass) and an increase in *Calluna* (heather), which indicate that *Calluna* now for the first time started to spread in pastures, probably as a result of high grazing pressure.

Zone 4 – Abandonment and reforestation (c. A.D. 1350–1550):
The many changes in the upper part of the diagram from Grisavad make the distinction of zones problematic. Zone 4, for example, could have been divided into two or three zones, but as
the changes are interpreted as a succession they are more easily described together. The beginning of the zone, dated to c.
A.D. 1350, is characterized by a decrease in *Calluna* frequencies and a sharp increase in *Salix* (willow). The *Salix* peak is
then followed by a *Betula* (birch) peak, which is followed by a
sharp peak in *Fagus* (beech). This sequence of events reflects
that some cultural land was abandoned in c. A.D. 1350, and
that it was occupied by *Salix*, which was later shaded out by
Betula, which in turn was shaded out by *Fagus*. However, there
was no overall reforestation and parts of the landscape were
still kept open by grazing. Most interesting, however, is the absence of cereal pollen from this period. The abrupt end of the
cereal record by the beginning of zone 4 indicates that the local farm reflected in zone 3 was now abandoned. The area was
still used for some grazing by distant farms.

Zone 5 – Re-establishment of a local farm (c. A.D. 1550–1800):
The beginning of zone 5, dated to c. A.D. 1550, is characterized
by a decrease in *Fagus* (beech) and an increase in *Secale* (rye)
and Cerealia undiff. (unspec. cereals), indicating that the *Fagus*
woodland that expanded on abandoned cultural land during
the previous period was now cleared and cultivation re-introduced. This initial clearing is followed by a very sharp increase
in Poaceae undiff. (grass) and the appearance of several cultivated taxa. The signal of cultivation is strong throughout this
period, which certainly reflects the arable fields of the croft next
to the peatland that were archaeologically documented (see
above). Cultivated taxa in this zone and the succeeding zone 6
are Cerealia undiff. (unspec. cereals), *Hordeum* type (barley),
Secale (rye), *Triticum* (wheat), *Avena* (oats), *Fagopyrum esculentum* (buckwheat), and *Linum usitatissimum* (flax).
    The grasslands (probably both pastures and meadows), reflec-
ted in for example Poaceae undiff., *Calluna* (heather), *Juniperus*

(juniper), *Plantago lanceolata* (ribwort plantain), etc., are more complicated to interpret. The very high values of Poaceae undiff. in the beginning of zone 5 soon decline. They are followed by peaks in *Myrica* (bog-myrtle), *Calluna*, *Alnus* (alder) and *Betula*, which in turn are followed by peaks in *Fagus*, *Pinus* (pine) and *Quercus* (oak). Obviously some grassland was abandoned soon after the initial expansion. The peaks in *Myrica* and *Alnus* show that some grassland on wet soils was overgrown, probably on the peatland, but the increase in *Fagus* and *Quercus* shows that some grassland on dry ground was also abandoned. The reforestation only affected parts of the landscape, and there were still large parts that were kept open by grazing and other agricultural land-use. The cereal record is continuous so the local farm – the croft – was not abandoned.

Zone 6 – Heathland and local farm (c. A.D. 1800–1900): This zone reflects first of all *Calluna* heathland. Pollen frequencies of *Calluna* (heather) increase greatly at the beginning of the zone at the same levels where *Pinus* (pine), *Fagus* (beech) and *Quercus* (oak) decrease. The increase of *Calluna* could to some degree reflect that pastures changed their floristic composition, i.e. they were transformed from grasslands to heathland, due to land-use changes. Such changes may have been, for example, increased grazing pressure and/or the introduction of fire clearing. However, the values for Poaceae undiff. (grass) do not decrease, which, together with the decrease of *Pinus*, *Fagus* and *Quercus*, show that there was also a pronounced total expansion of grazing land. The very sharp decrease in *Pinus* shows that the *Pinus* stands that were established on abandoned pastures in the end of zone 5 were now cleared and replaced by *Calluna* heathland. The strong indications of cultivation that were recorded in zone 5, and which were connected to a local farm, continue throughout zone 6.

Zone 7 – Abandonment and reforestation (c. A.D. 1900–present): The beginning of zone 7 is characterized by a sharp decrease in *Calluna* (heather) and the termination of cultivation indicators, together with an increase in *Betula* (birch), *Pinus* (pine), and *Picea* (spruce). These changes reflect the abandonment of the local croft and its agricultural land as well as the surrounding heathland, and in the same process the introduction of coniferous plantation, that is the beginning of modern silviculture.

Värsjö Utmark

The sampling site Värsjö Utmark is situated in the northeastern part of the investigation area, on a plateau of sandy till and peatlands. The area is almost completely forested, mainly with planted *Picea* (spruce) and *Pinus* (pine), but also with some small pastures. It has a rather flat or gently undulating topography and an altitude of 120–130 m above sea level. About one kilometre to the west runs the small river Pinnån in a narrow valley with glaciofluvial deposits.

The sampled peatland is a small bog with tree vegetation of spruce (*Picea abies*) and pine (*Pinus sylvestris*). It is drained by a ditch, dug in connection to silviculture, which has resulted in secondary compaction (but not destruction) of the uppermost peat layers. The coring point was situated approximately 20 m from the edge of the peatland, facing an archaeological investigation site which was subject to detailed mapping and excavation (figure 14).

Most of the documented features on the site could be connected to a croft from the 19th century. These features were stone walls, clearance cairns, field terraces, and the remains of at least three different buildings. One building, interpreted as some kind of outhouse, was partly excavated (remember that this type of excavation is restricted to the exploitation area), and yielded finds of pottery, glass, buttons, iron nails, etc (A. Knarrström 2004). The documented stone walls have enclosed an infield area with cultivated fields and hay meadows. Two clearance cairns were excavated and dated by radiocarbon analysis of in-mixed charcoal. One of them could be connected to the 19th century croft, while the other turned out to be from the

12th century, i.e. medieval (charcoal data are presented in more detail in a separate section in this appendix). The dated charcoal was of *Betula* (birch) in the late clearance cairn and of *Fagus* (beech) in the medieval one. Outside the infields to the southwest were the remains of charcoal kilns and tar pits, which in the same way as the clearance cairns could be referred to two separate periods of usage. The Middle Ages were represented by a kiln dated to the 13th century (*Fagus* and *Pinus* charcoal) and some tar pits of which one was dated to the 13–14th centuries (*Pinus* charcoal). A later period was represented by four kilns dated to the 18–19th centuries (*Pinus* and *Betula* charcoal). To sum up, the archaeological investigation has identified two periods of human activity on the site: one medieval period with dates from the 12th to the 14th centuries, and one post-medieval or modern with dates mainly from the 19th century but possibly the beginning of the 18th. From both periods there is evidence of agriculture as well as the use of woodlands for charcoal and/or tar production. The investigated building remains refer to the late period but also during the medieval period there must have been settlements somewhere in the vicinity.

In addition to the archaeological information from the Värsjö Utmark site, the results from a site called Bredabäck, situated only 800 m to the southwest, should also be mentioned (Forenius et al. 2005, Strömberg in prep.). At Bredabäck the remains of an iron production site were investigated, and charcoal from slag and furnaces was radiocarbon-dated. The results show that the production site was in use during the early Middle Ages, more precisely during the 13th–14th centuries, i.e. during the same time as the early settlement phase at Värsjö Utmark. Thus we can conclude that not only agriculture but also iron production was practised in the area during the early Middle Ages.

Back to Värsjö Utmark, the coring at the sampling point for pollen analysis revealed a total gyttja and peat thickness of 200 cm (table 9). Based on a reconnaissance analysis and three radiocarbon dates (Björkman & Ekström 2002), the sequence 0–65 cm below the surface was chosen for the detailed pollen

Table 9. Peat stratigraphy at the coring point at Värsjö Ut-mark. The sequence 0–65 was subject to pollen analysis.

Depth (cm)	Description
0–105	Fen peat with wood
105–190	Fen peat
190–200	Coarse detritus gyttja

Table 10. Radiocarbon dates from the peat sequence at Värsjö Utmark.

Depth (cm)	Lab.no	C14 BP	Cal 2s (max–min)	Dating material
6–7.2	Poz-7540	215 ± 30	A.D. 1640–1950	Pollen concentrate
10.7–12	Poz-7539	340 ± 30	A.D. 1470–1640	Pollen concentrate
15.5–16.5	Poz-7538	945 ± 30	A.D. 1020–1170	Pollen concentrate
20–21	Ua-26404	1150 ± 45	A.D. 770–990	Fen peat cleared from roots and wood
60–61	Ua-26405	2145 ± 50	360–40 B.C.	Fen peat cleared from roots and wood
149–150	Ua-26406	2715 ± 50	980–740 B.C.	Fen peat cleared from roots and wood

analysis presented in this book. Due to the secondary compaction, the sequence had to be subsampled very densely, in particular in the upper part. The analysed material was fen peat with some wood. From the sequence 0–65 cm five levels were radiocarbon-dated (table 10).

The original pollen analysis covered the last approximate 2200 years, while the diagram presented here covers the last 2000 years (figure 51). It has been divided into seven zones which will be briefly described. These zones are pollen assemblage zones (Hedberg 1976, Berglund & Ralska-Jasiewiczowa 1986), but with emphasis on human impact indicators. As the diagram is plotted on a linear time scale the zones are sometimes also referred to as periods.

Zone 1 – Wood pasturage (>0 B.C.–c. A.D. 850): This zone reflects a forested landscape with *Betula* (birch) and *Alnus* (alder) growing on the peatland and *Quercus* (oak), *Tilia* (lime) and *Corylus* (hazel) dominating on dry ground. The period from c. A.D. 300 onwards is distinguished as subzone 1b, characterized by slowly increasing *Fagus* (beech) values. In the upper part of 1b there is also a decrease in *Corylus*, which may be matched with a gradual increase in *Betula*.

Throughout zone 1 there is a weak but regular occurrence of herb taxa, for example an almost continuous curve for *Plantago lanceolata* (ribwort plantain), which indicates grazing. The high frequencies of tree pollen and the low frequencies of for example Poaceae undiff. (grass) indicate, however, that the grazing only resulted in minor openings in the forest.

Zone 2 – Temporary cultivation and wood pasturage (c. A.D. 850–1150): The gradual increase in *Fagus* (beech), which started in the previous zone, continues throughout zone 2 and reaches maximum values by the end of the zone, dated to c. A.D. 1100. By this time *Fagus* was the dominating shade-tolerant tree on dry ground in the area. *Quercus* (oak) and *Tilia* (lime) were partly outcompeted by *Fagus*, although *Quercus* was still rather common.

The weak but continuous grazing-signal of zone 1 continues throughout zone 2, indicating continuity in extensive wood pasturage. A few pollen grains of *Secale* (rye) show that some small-scale cereal growing was introduced, probably as slash-and-burn cultivation or some other kind of temporary cultivation. A peak in microscopic charcoal may indicate the use of fire in these clearings.

Zone 3 – Permanent cultivation by local farm (c. A.D. 1150–1350): The beginning of zone 3 marks deforestation and the establishment of a relatively open cultural landscape. Decreasing *Fagus* and *Quercus* values reflect the clearing of beech and oak forest to create open land for pastures, meadows and cultivated fields. (This picture of the early-medieval deforestation in the area, which is based on pollen data, fits in well with the finding of macroscopic *Fagus* charcoal beneath a medieval clearance cairn adjacent to the peatland; see above.) The high frequencies of grass pollen and a large number of herb taxa, together with the rather low frequencies of *Calluna* (heather), reflect pastures with rich grassland vegetation. Microscopic charcoal shows a strong peak in the lowermost sample but low values throughout the rest of the zone, which indicates that fire was used in the initial clearings but not in the succeeding land-use.

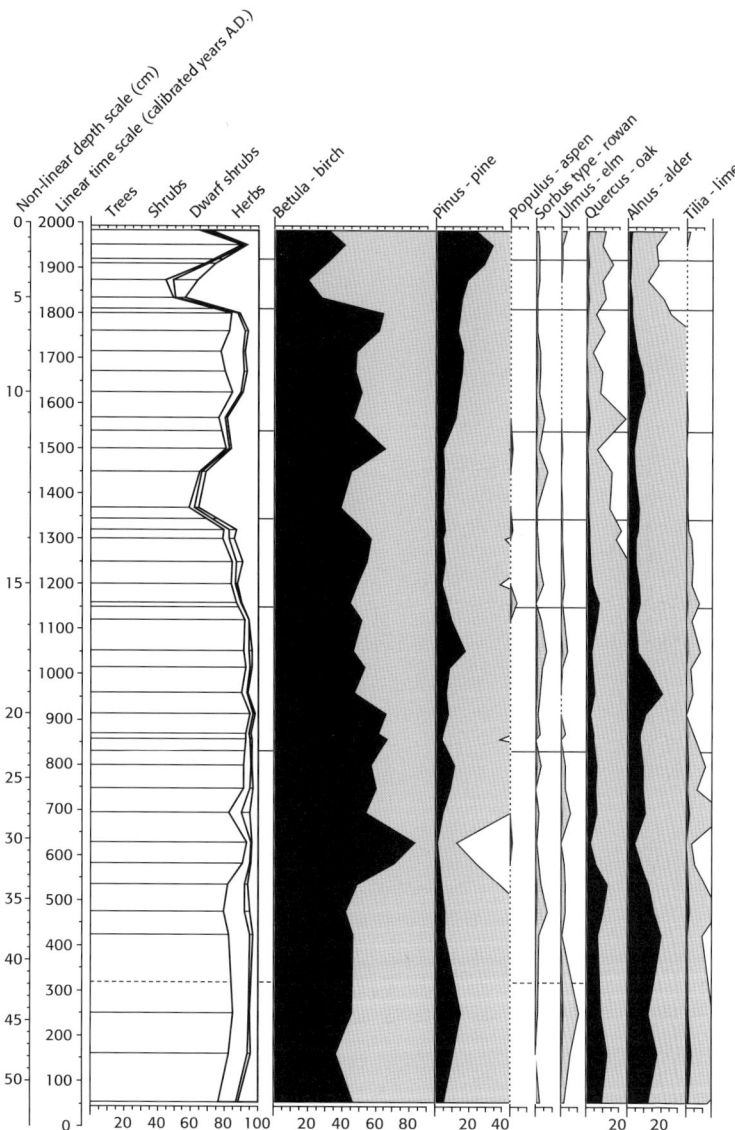

Fig. 51. Total pollen diagram from Värsjö Utmark (56°19'N, 13°26E) with all identified pollen taxa. The graphs show pollen percentages (black) and ten times exaggeration (shaded), and they are based on the total pollen sum. Taxa presented to the left of the pollen sum in the diagram are included in the sum. Analysed by Nils-Olof Svensson and Per Lagerås.

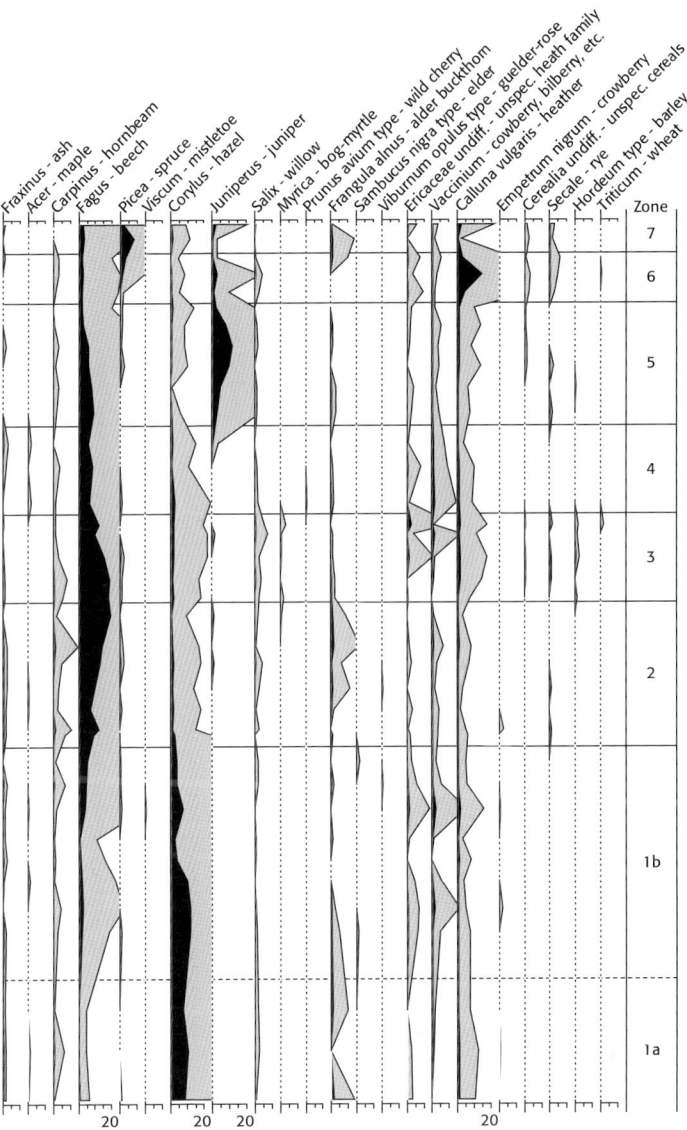

Fig. 51. Continued.

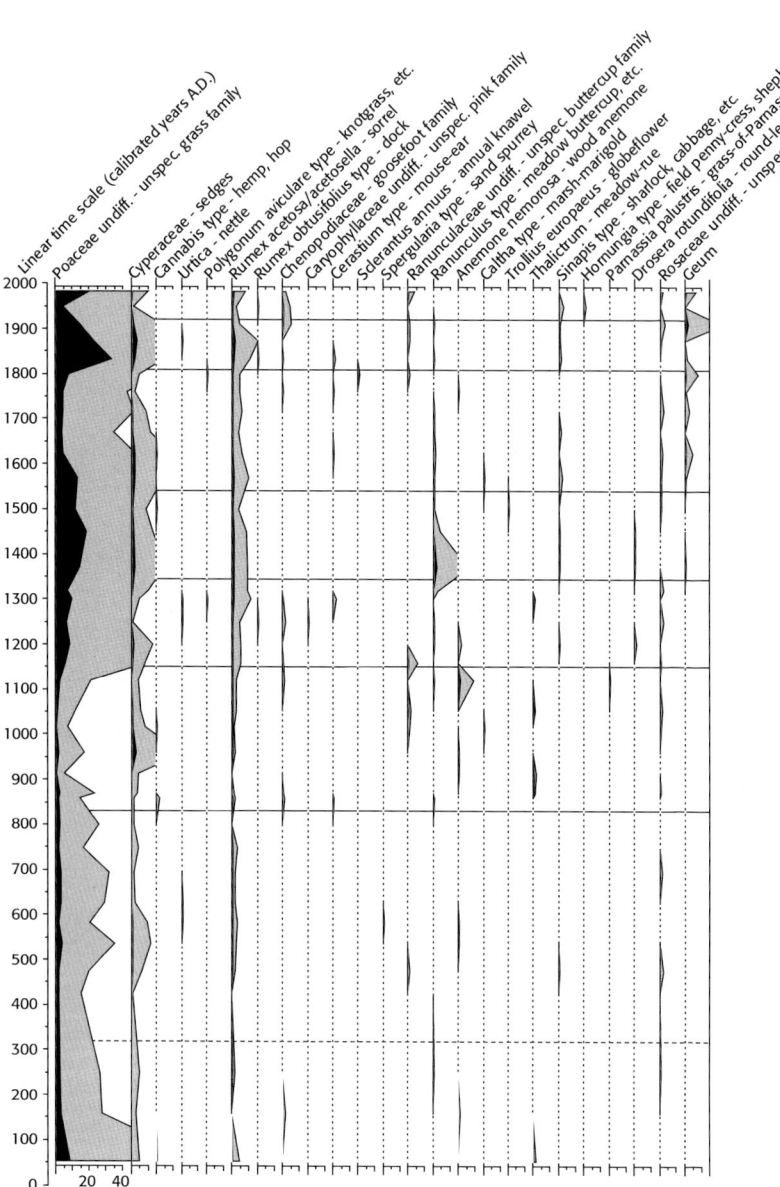

Fig. 51. Continued.

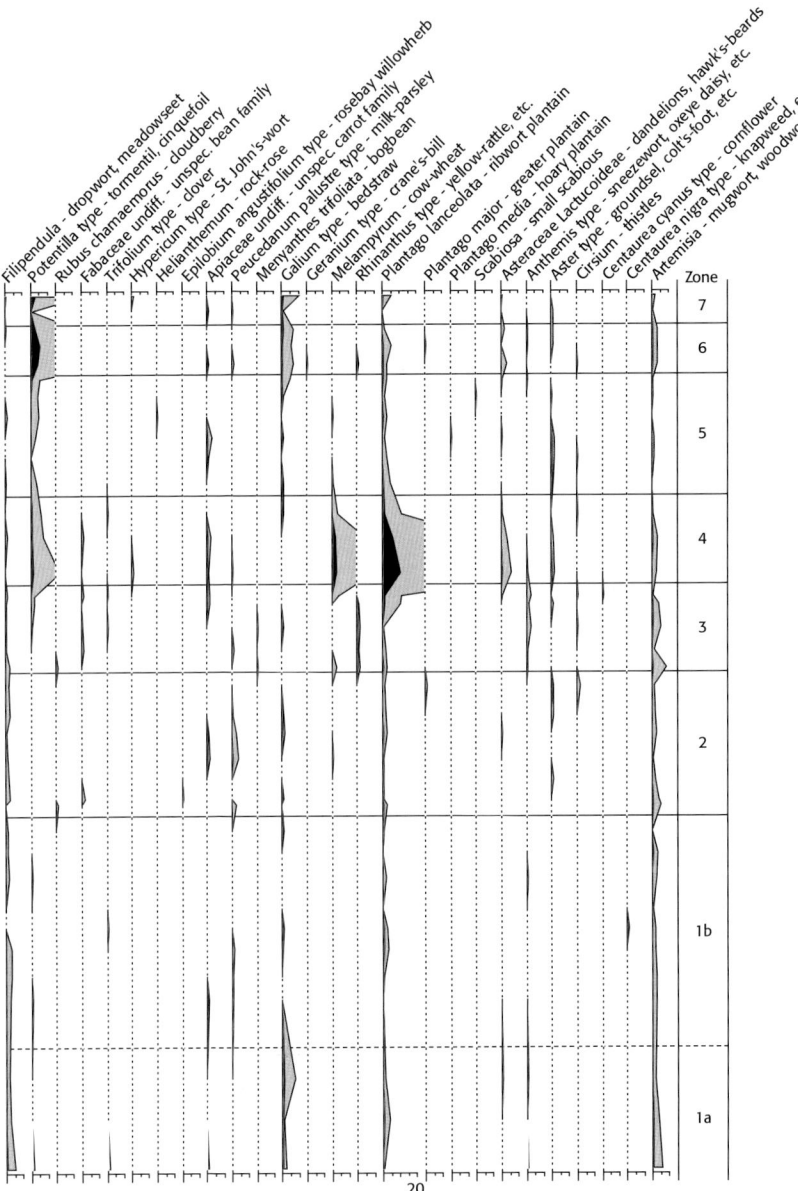

Fig. 51. Continued.

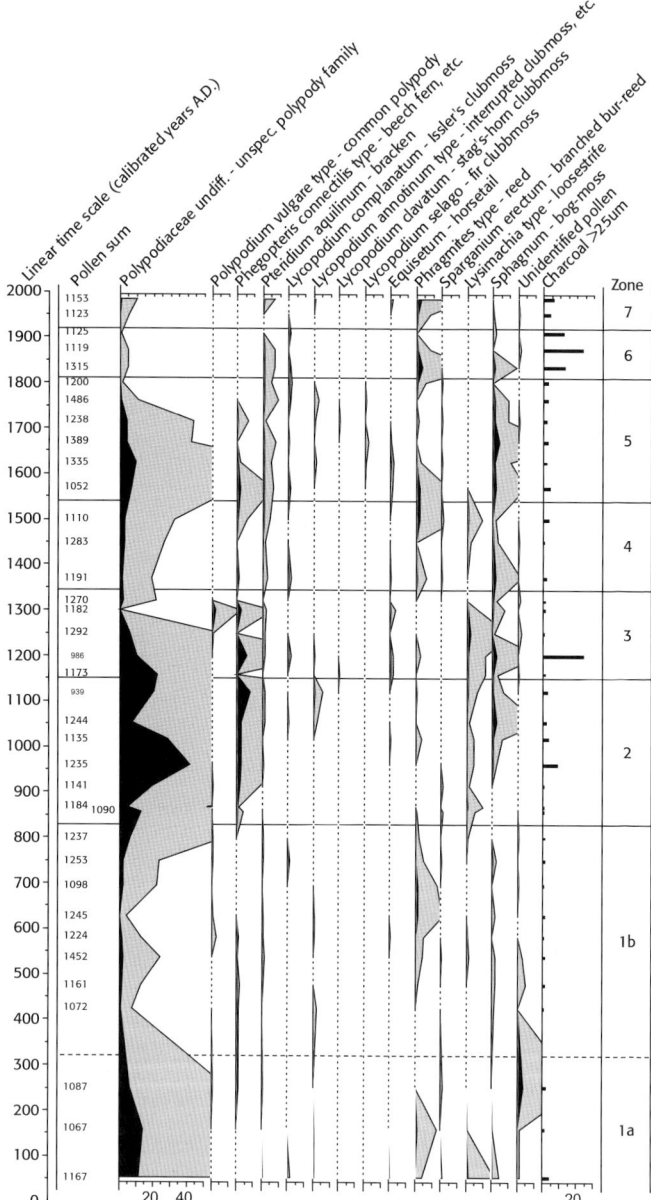

Fig. 51. Continued.

Apart from grassland indicators this zone is characterized by a continuous record of cereal pollen, which indicates permanent cultivation in connection to a local farm. Cultivated taxa were *Secale* (rye), *Hordeum* (barley), and *Triticum* (wheat). The agricultural activity reflected in zone 3 can be connected to the early-medieval remains that were discovered and documented during the archaeological excavation next to the peatland, and during the same period there was also iron production going on at the nearby site Bredabäck (see above).

Zone 4 – Farm abandonment and grazing (c. A.D. 1350–1550): The beginning of zone 4 is characterized by an abrupt end of the cereal record, which indicates that the farm of the previous period had been abandoned. Notable is also the absence of some other pollen taxa that indicate arable fields or ruderal land, and which were recorded in the previous zone, such as *Urtica* (nettle), *Polygonum aviculare* type (knotgrass), *Rumex obtusifolius* type (dock), Chenopodiaceae (goosefoot family), etc. The absence of these taxa in zone 4 support the picture given by the cereal record that arable fields and settlement were now deserted. However, the area was still used for grazing, as indicated by the high frequencies of Poaceae undiff. (grass), *Plantago lanceolata* (ribwort plantain), and several other grassland indicators. The pollen values for these grassland taxa are even higher than in the previous zone, which may indicate an expansion of grasslands (or possibly a decrease in grazing pressure that resulted in enhanced flowering). The plant composition also changed, which may reflect a change in grassland use. *Rhinanthus* type (yellow-rattle), for instance, shows a continuous record throughout zone 3 but is absent in zone 4. This may indicate that grasslands previously used for hay mowing had been turned into pastures.

Zone 5 – Shrubs and cultivation (c. A.D. 1550–1800): At the beginning of zone 5 there is a strong increase in *Juniperus* (juniper) pollen. Together with a decrease in Poaceae undiff. (grass) the high *Juniperus* frequencies indicate abandonment and overgrowing of pastures. In the upper part of the zone *Juniperus* decreases while *Betula* increases, which probably reflects a succession in which *Juniperus* shrubs were finally outshaded by *Betula*. These changes together indicate decreased grazing pressure and abandonment of some pastures in the beginning of zone 5, that is in the mid-16th century. At the same levels, however, cereal pollen appears which may seem a contradiction. The most probable interpretation is that after some hundred years of only grazing during period 4, a farm had been established somewhere in the vicinity. In connection with this, cereal cultivation and probably hay mowing was re-introduced. Grazing was reorganized in a way that resulted in the overgrowing of some pastures, in particular close to the sampled peatland.

Zone 6 – Heathland and local farm (c. A.D. 1800–1900): The beginning of zone 6, which is dated relatively uncertainly, is characterized by a very sharp decrease in *Betula* (birch) and an equally sharp increase in Poaceae undiff. (grass). Other cultural indicators also increase, such as for example *Rumex acetosa/acetosella* (sorrel), *Potentilla* type (tormentil, cinquefoil), *Galium* type (bedstraw), and *Artemisia* (mugwort, wormwood). It reflects how *Betula* woodland was cleared and replaced by open pastures and arable fields. (The interpretation is supported by the identification of macroscopic *Betula* charcoal found beneath a clearance cairn tentatively dated to this period; see above.) As the *Betula* vegetation was the result of the abandonment and overgrowing of previously open land, we can conclude that the land that was cleared in the beginning of period 6 was the same land that had been open during periods 3 and 4.

Another interesting characteristic of zone 6 is that the sharp peak in Poaceae undiff. at the beginning of the zone further up is replaced by a peak in *Calluna vulgaris* (heather). This change reflects how the open grasslands that were once created by the clearing of *Betula* woodlands soon turned into *Calluna* heathland. The high frequencies of microscopic charcoal throughout this period probably originate from the practice of regular and intentional burning to rejuvenate heather.

Apart from pastures, the pollen diagram also proves the existence of arable fields during period 6, reflected in *Secale* (rye), *Triticum* (wheat) and Cerealia undiff. (unspec. cereals) pollen. The agricultural activity reflected in zone 6 can certainly be connected with the archaeological remains of a 19th century croft that were excavated next to the peatland.

Zone 7 – Abandonment and reforestation (c. A.D. 1900–present): The beginning of zone 7 is characterized by a decrease in *Calluna* (heather) and an increase in *Betula* (birch), *Pinus* (pine), and *Picea* (spruce), which reflect the abandonment of heathland and other agricultural land and the establishment of coniferous forest, i.e. the beginning of modern silviculture. This reflects the present day landscape, in which there is no local farm and the area is used for silviculture and some grazing.

Bjärabygget

The sampling site Bjärabygget is situated in the north-easternmost part of the investigation area, 1 km south of Skånes-Fagerhult and about 5 km northeast of Värsjö Utmark presented above. Only six kilometres to the north runs the border between the provinces of Skåne and Småland, which until 1658 was the national border between Denmark and Sweden. At Bjärabygget there is a gently undulating landscape of sandy till and peatland, and the area is situated at an altitude of 120–130 m above sea level. The landscape is completely forested, planted with pine (*Pinus sylvestris*) and spruce (*Picea abies*).

The investigation site was a south-facing slope and a plateau bordering a relatively large peatland. On dry ground were the visible remains of ancient agriculture and settlement, such as clearance cairns, field terraces, and stone foundations of buildings. The site was subject to an archaeological excavation within the project (Linderoth 2004), during which the agricultural remains were documented and sampled for radiocarbon dating (Lorentzon in prep.). The habitation area with building remains could however not be excavated as it was situated outside the area for exploitation. Among the agricultural remains sparse finds were made of for example pottery, glass, china, buttons, iron nails, and some iron tools. During the excavation a small Neolithic hunting station close to the peatland was also documented but that is outside the scope of this book (for a thorough presentation of the Stone Age in the area see B. Knarrström 2007).

Looking in more detail at the agricultural remains, altogether ten macroscopic charcoal pieces from eight different clearance cairns were dated, and the results are presented in the diagram in figure 17. As evident from the diagram two charcoal pieces were dated to the 10th century, i.e. the Viking Period, six were dated to a period which lasted from the 13th to the 15th century, i.e. the Middle Ages, and finally two (both from the same clearance cairn) were dated to the 18th or 19th century. Of the two early dates from the Viking Period, one is from a clearance cairn that also gave a date to the late Middle Ages. Probably the older date is not connected to stone clearing but may still reflect clearing by fire for agricultural purposes. To conclude, the radiocarbon dates from Bjärabygget show that most of the dated clearance cairns originate in medieval cultivation, or more precisely from the 13th to the 15th century. This period of cultivation was preceded by some temporary clearings some centuries earlier, during the Viking Period. One clearance cairn, finally, was dated to the 18th–19th century and thus reflects the latest phase of cultivation on the site.

On the oldest detailed map of the area, dated to 1714, the investigation site is situated in the outland between the two hamlets Fagerhult and Bjärabygget. The map shows no settlement on the site, but on a later map, dated to 1818–1822, there are two houses and some in-fenced areas for cultivation or grazing. A map from 1928 shows no houses on the site but there was still a small enclosed pasture. It is interesting to note that the map from 1714 names the site *Twiste Platz*, which means "a place of dispute" and probably indicates that there were different opinions as to if this place belonged to Fagerhult or Bjärabygget. Obviously this small piece of outland had some value, even though there was no settlement at the time. As the site had once been settled, cultivated and stone cleared during the Middle Ages, it may have been regarded as a valuable resource which could be taken into use, and therefore was worth disputing (Lorentzon in prep.).

As mentioned above, the clearance cairns and other remains on the site bordered a peatland. The central parts of this large bog is characterized by hare's-tail cottongrass (*Eriophorum vaginatum*) and scattered dwarf trees of pine (*Pinus sylvestris*), while the marginal fen also has sedges (*Carex* sp.), bog asphodel (*Narthecium ossifragum*), Labrador-tea (*Ledum palustre*), etc., as well as denser tree stands of pine and birch (*Betula* sp.). A coring far out on the bog revealed a total peat and gyttja thickness of five metres, reflecting a lake-to-fen transition probably as early as the Stone Age. For the pollen analysis a marginal coring point was chosen in close connection to the archaeological excavation site, approximately 30 metres from dry ground and the nearest clearance cairn. It revealed a total peat and gyttja stratigraphy of 1.85 m, dominated by fen peat (table 11), of which the sequence 0–96 cm was subject to a detailed pollen analysis. Four levels were radiocarbon-dated (table 12).

The original pollen analysis covered the last 2300 years, while the diagram presented here covers the last 2000 years (figure 52). It has been divided into six zones which will be briefly described. These zones are pollen assemblage zones (Hedberg 1976, Berglund & Ralska-Jasiewiczowa 1986), but with emphasis on human impact indicators. As the diagram is plotted on a linear time scale the zones are sometimes also referred to as periods in the text.

Table 11. Peat stratigraphy at the coring point at Bjära-bygget. The sequence 0–96 cm was subject to a detailed pollen analysis.

Depth (cm)	Description
0–4	Moor
4–20	Fen peat, moderately humified
20–40	Fen peat, strongly humified
40–52	Carex peat with wood
52–160	Fen peat with wood
160–185	Coarse detritus gyttja, sandy in the lower part

Table 12. Radiocarbon dates from the analysed peat sequence at Bjärabygget.

Depth (cm)	Lab.no	C14 BP	Cal 2s (max–min)	Dating material
20–21	LuS-6209	370 ± 50	A.D. 1440–1640	slender, charred twigs
30–31	LuS-6208	785 ± 60	A.D. 1040–1390	fen peat cleared from roots and wood
42–43	LuS-6207	1055 ± 50	A.D. 870–1150	Carex peat cleared from roots and wood
70–72	LuS-6206	1720 ± 50	A.D. 160–430	terrestrial macrofossils (mainly insects)

Zone 1 – Wood pasturage (>0 B.C.–c. A.D. 900): Zone 1 reflects a forested landscape with *Betula* (birch) and some *Alnus* (alder) growing on the peatland and *Quercus* (oak), *Tilia* (lime) and *Corylus* (hazel) dominating on dry ground. The period from c. A.D. 350 onwards is distinguished as subzone 1b, characterized by increasing *Fagus* (beech) values and slowly decreasing *Corylus* values.

Throughout zone 1 there is a weak but regular occurrence of herb taxa, for example in an almost continuous curve for *Plantago lanceolata* (ribwort plantain), which indicates grazing. There are also relatively high frequencies of Poaceae undiff. (grass), which may indicate pastures on dry ground but more probably at least to some degree reflect grass vegetation on the fen.

Zone 2 – Temporary cultivation (c. A.D. 900–1250): The transition from zone 1 to zone 2 is characterized by a short but sharp increase in *Betula* (birch) in the upper part of zone 1, which in the beginning of zone 2 is followed by the appearance of *Secale* (rye) and a strong increase in *Fagus* (beech). This series of changes probably reflects clearings for temporary cultivation – a disturbance that favoured the local expansion of beech (cf. Björkman 1996a; see also zone 1 at Grisavad). The weak and discontinuous cereal record together with the low frequencies of, for example, *Rumex acetosa/acetosella* (sorrel) indicate only temporary cultivation, probably by a farm at some distance. Two of the analysed charcoal pieces from the archaeological excavation were dated to the 10th century, and may be connected with the temporary clearings and cultivation reflected in the pollen diagram.

Zone 3 – Permanent cultivation by local farm (c. A.D. 1250–
1700): From the beginning of zone 3, dated to c. A.D. 1250,
there is a continuous record of *Secale* (rye) and also some pollen of Cerealia undiff. (unspec. cereals). Together with the relatively high frequencies of *Rumex acetosa/acetosella* (sorrel)
these cereals indicate permanent cultivation by a local farm.
The zone is also characterized by relatively high values for
Poaceae undiff. (grass) which indicate pastures and possibly
hay meadows. Throughout zone 3 there is a gradual decrease
in *Fagus* (beech) percentages and a gradual increase in *Calluna*
(heather), which reflects the long-term effect of grazing.

The interpretation of local cultivation at the site during the
Middle Ages is supported by a series of radiocarbon-dated
clearance cairns next to the peatland. Six of the clearance
cairns were dated to the time period represented by zone 3.
Note in particular that the continuous series of dated clearance
cairns start at the same time as the continuous record of cereals in the pollen diagram. Hence there is a striking similarity
between these two independent records, showing the establishment of permanent cultivation in the 13th century.

The upper part of zone 3, dated to c. A.D. 1550–1700, is
distinguished as subzone 3b. The beginning of the subzone is
characterized by an increase in cereal pollen, indicating intensified cultivation. However, further up the subzone is characterized by a decrease in grazing indicators such as *Calluna* and
Poaceae undiff. and also in microscopic charcoal, and an increase in *Betula* (birch). It reflects abandonment and overgrowing of some pastures. However, as the cereal record shows continuity throughout the subzone this change is not interpreted
as farm abandonment, but rather as a change in grazing regime
and pasture management. Possibly a cease in the use of fire (to
rejuvenate heather) resulted in overgrowing.

Zone 4 – Heathland (c. A.D. 1700–1900): The beginning of
zone 4 is characterized by a sharp decrease in *Betula* (birch), and
a sharp increase in *Calluna* (heather). At the same levels there is
also a sharp, although slightly delayed, increase in *Pinus* (pine).

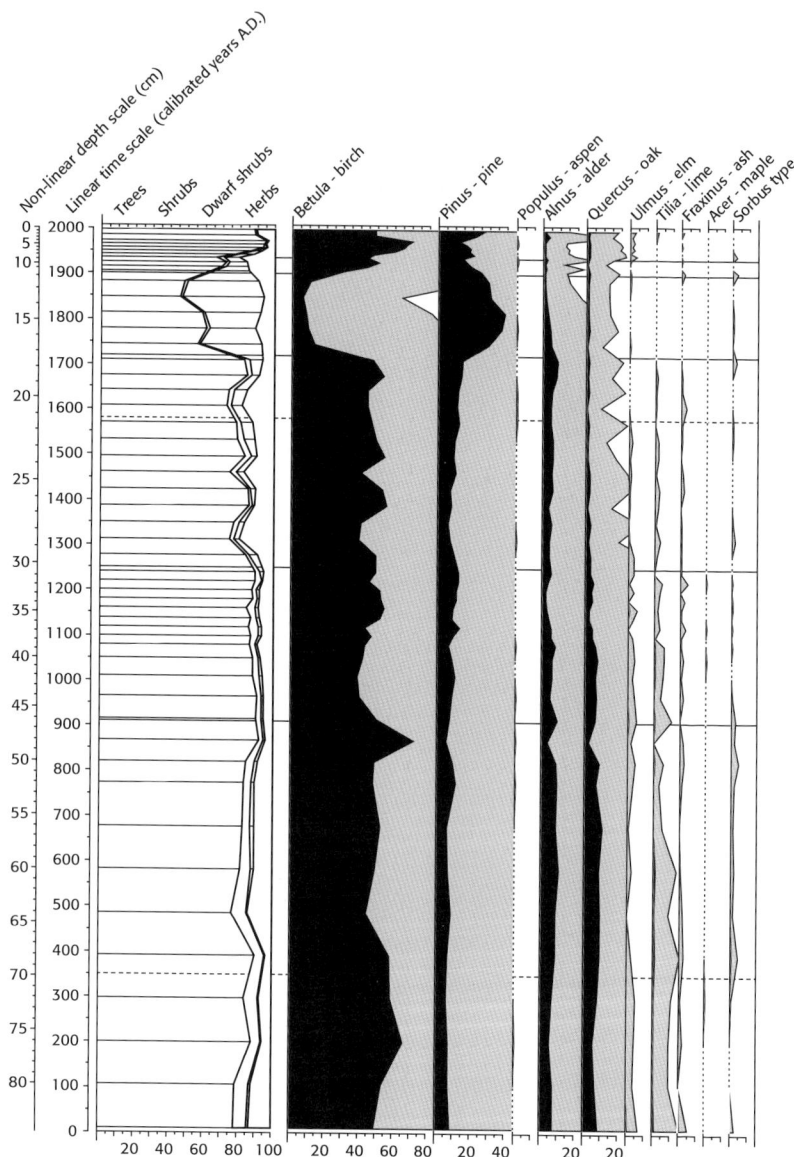

Fig. 52. Total pollen diagram from Bjärabygget (56°21'N, 13°28E) with all identified pollen taxa. The graphs show pollen percentages (black) and ten times exaggeration (shaded), and they are based on the total pollen sum. Taxa presented to the left of the pollen sum in the diagram are included in the sum. Analysed by Nils-Olof Svensson and Per Lagerås.

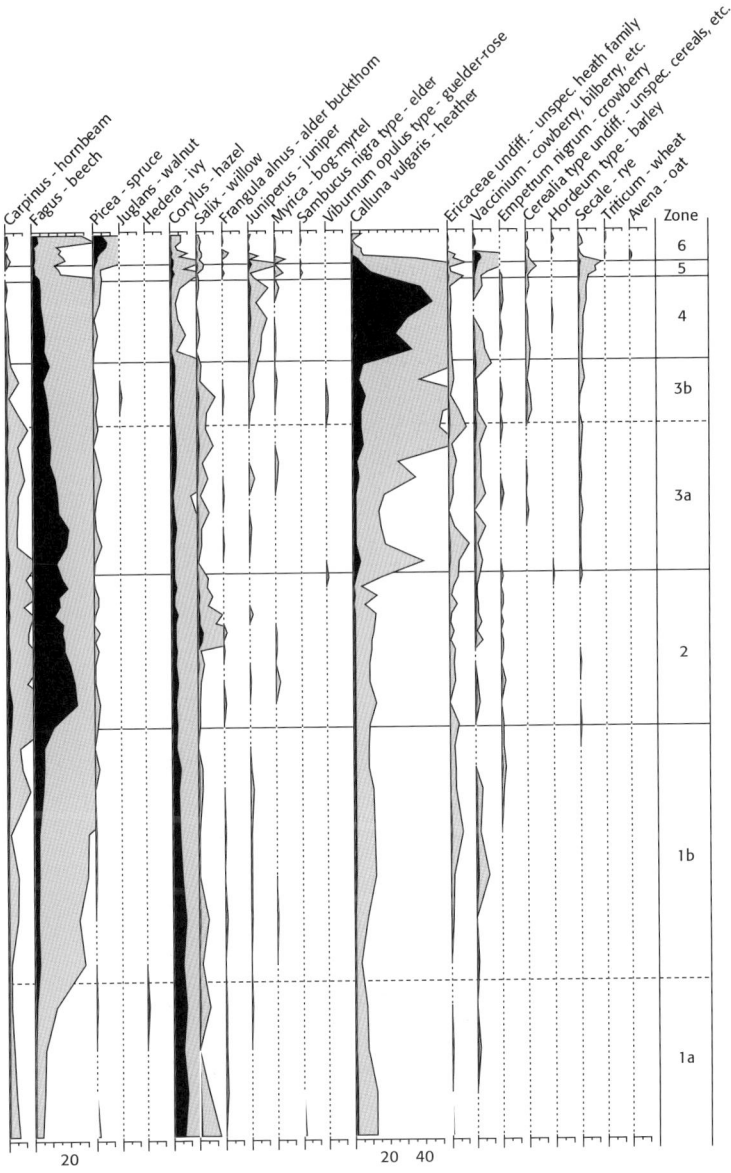

Fig. 52. Continued.

APPENDIX 2

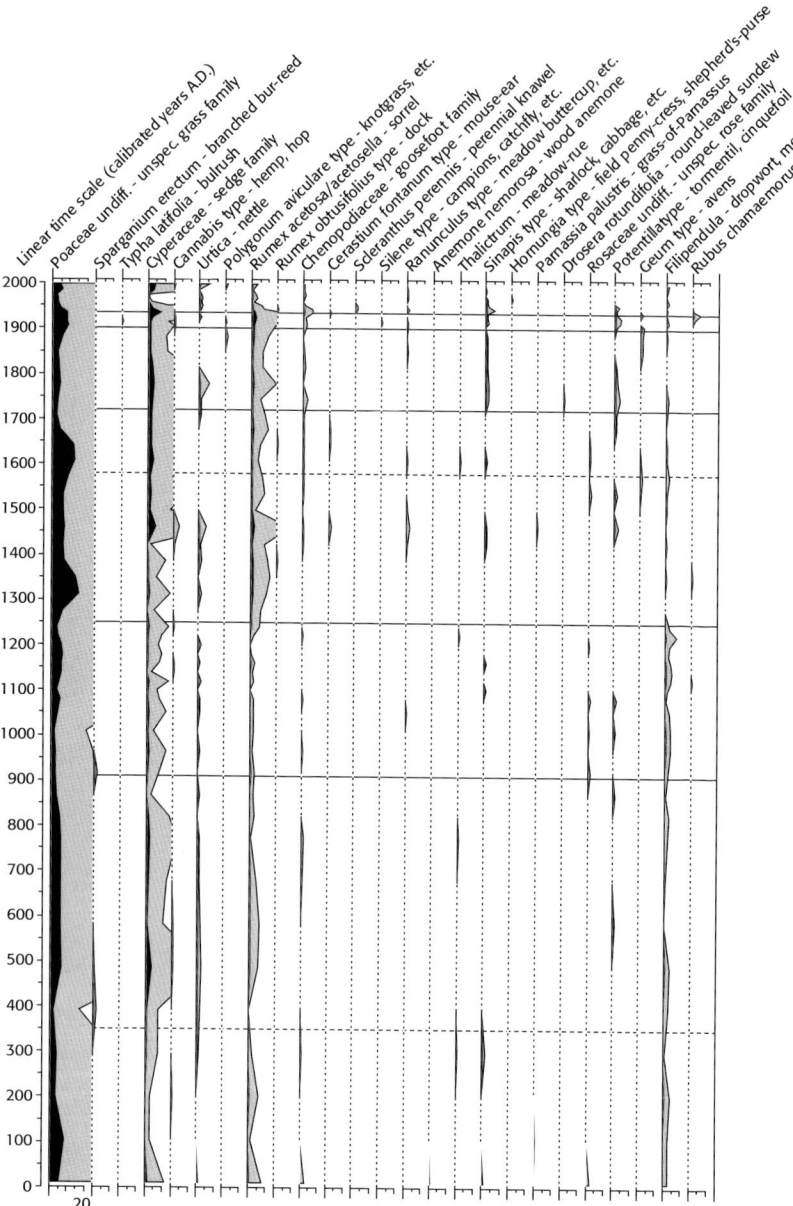

Fig. 52. Continued.

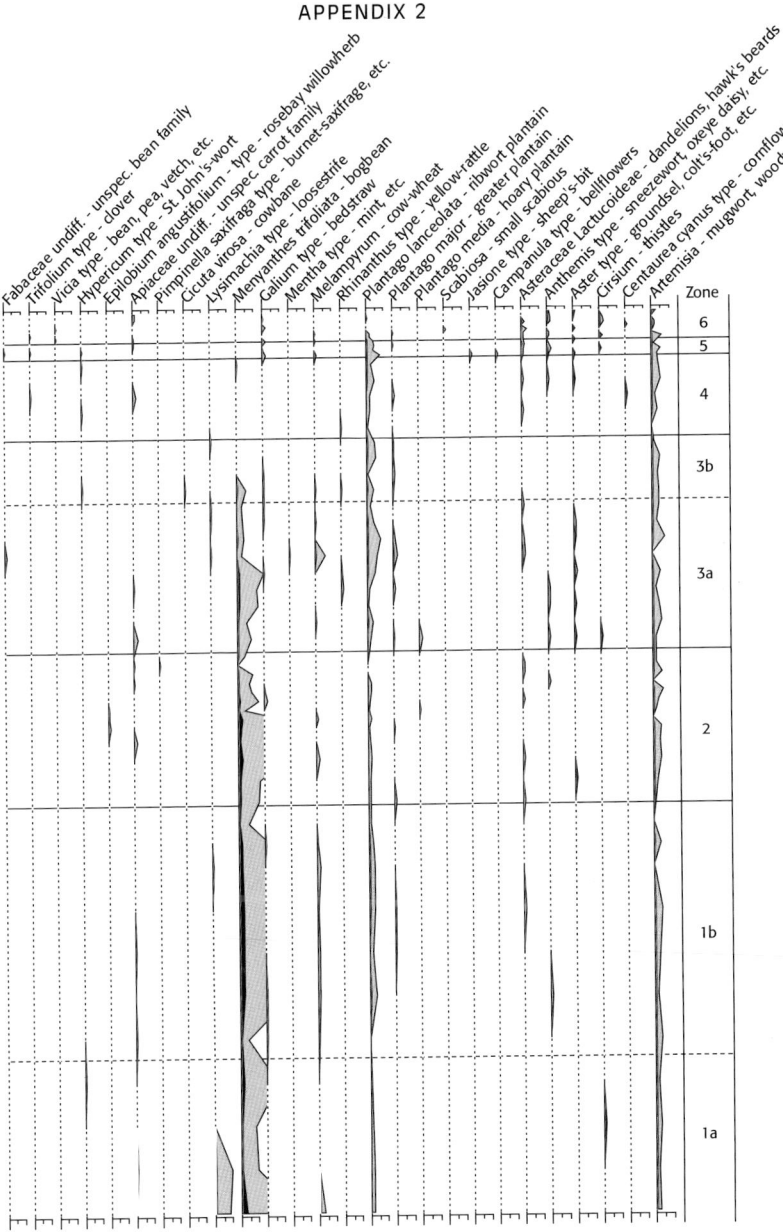

Fig. 52. Continued.

APPENDIX 2

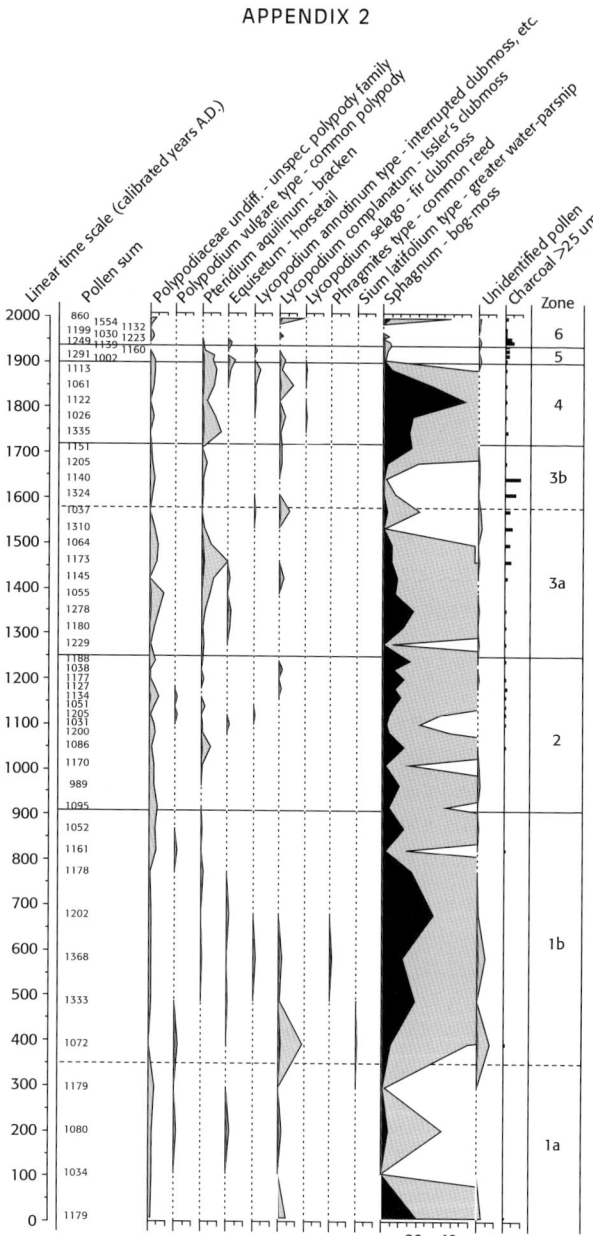

Fig. 52. Continued.

Obviously, almost all *Betula* woodland in the vicinity was cut down and replaced by heathland with scattered pine vegetation. The very low frequencies of microscopic charcoal show that fire played no role in the woodland clearings or in the succeeding management of the heathland. In this respect the diagram from Bjärabygget differs from the other three local pollen diagrams from the project, in all of which the high *Calluna* values in the upper parts of the diagrams are matched by high frequencies of microscopic charcoal. There is also a difference in timing. At the other sites the high *Calluna* values are by and large confined to the 19th century, while at Bjärabygget the major *Calluna* expansion starts as early as the early 18th century.

Throughout zone 4 there is a continuous cereal record, in particular of *Secale* (rye), which shows that cultivation continued, possibly by the same farm as in zone 3. The cereal values are even slightly higher than in zone 3, but that may be due to increased pollen transport following the deforestation.

Zone 5 – Cultivation and grassland (c. A.D. 1900–1950): In zone 5, which approximately represents the early 20th century, *Calluna* (heather) values decrease while *Betula* (birch) values increase. It reflects decreased grazing and reforestation on heathland. At the same time, however, cereal pollen reaches rather high values (maximum values in the diagram), and so does *Rumex acetosa/acetosella* (sorrel), indicating permanent cultivation. Also for instance Poaceae undiff. (grass) reaches relatively high values indicating pastures or hay meadows. The pollen record thus seems to be rather contradictory, with reforestation occurring at the same time as increased cultivation and grassland management, and it cannot easily be interpreted in terms of expansion or regression. However, it probably reflects a shift to more intensive land-use, in which extensive grazing on poor heathland was replaced by grazing on managed and more fertile grassland.

Zone 6 – Abandonment and reforestation (c. A.D. 1950–present): The uppermost zone in the diagram reflects farm abandonment and reforestation, mainly by *Picea* (spruce), i.e. the establishment of the present forested landscape.

Charcoal data

Macroscopic charcoal for analysis and radiocarbon dating was collected from archaeological features at a large number of excavation sites along the investigation area in northern Scania. Some of these data have been presented and discussed in earlier chapters in this book. Below is a list of all identified and radiocarbon-dated charcoal pieces from the project.

Table 13. Charcoal data from excavations within the project.

Charcoal taxa	Lab.no	C14 BP	Cal 2s max–min	Archaeological context
Alnus	Ua-26822	330 ± 30	A.D. 1480–1650	Post hole
Alnus	Ua-26818	820 ± 35	A.D. 1160–1290	Clearance cairn
Alnus	Ua-26893	894 ± 45	A.D. 1020–1250	Hearth
Alnus	Ua-27018	2895 ± 45	1260–920 B.C.	Hearth
Alnus	Ua-26998	2940 ± 40	1290–1000 B.C.	Cooking pit
Alnus	Ua-27006	5225 ± 45	4230–3950 B.C.	Hearth
Alnus	Ua-26756	5395 ± 40	4340–4040 B.C.	Hearth
Alnus	Ua-27008	5935 ± 45	4940–4710 B.C.	Cooking pit
Betula	Ua-26839	75 ± 30	A.D. 1680–1960	Clearance cairn
Betula	Ua-26961	80 ± 45	A.D. 1670–1960	Hearth
Betula	Ua-27016	100 ± 40	A.D. 1670–1960	Hearth
Betula	Ua-26952	265 ± 40	A.D. 1480–1950	Clearance cairn
Betula	Ua-26944	270 ± 45	A.D. 1480–1950	Clearance cairn
Betula	GrN-28687	280 ± 20	A.D. 1520–1670	Iron production site
Betula	Ua-26965	285 ± 45	A.D. 1480–1800	Hearth
Betula	Ua-26862	295 ± 35	A.D. 1480–1670	Charcoal production site
Betula	GrN-28685	300 ± 25	A.D. 1490–1660	Iron production site
Betula	GrN-28330	315 ± 25	A.D. 1480–1650	Iron production site
Betula	GrN-28333	315 ± 15	A.D. 1510–1650	Iron production site
Betula	GrN-28684	320 ± 20	A.D. 1490–1650	Iron production site
Betula	Ua-26753	325 ± 30	A.D. 1480–1650	Charcoal production site
Betula	Ua-26803	335 ± 40	A.D. 1460–1650	Clearance cairn
Betula	GrN-28688	365 ± 20	A.D. 1450–1630	Iron production site
Betula	GrN-28332	375 ± 20	A.D. 1440–1630	Charcoal production site
Betula	GrN-28686	375 ± 20	A.D. 1440–1630	Iron production site
Betula	GrN-28683	380 ± 20	A.D. 1440–1630	Iron production site
Betula	Ua-26826	405 ± 30	A.D. 1430–1630	Clearance cairn
Betula	Ua-26959	440 ± 60	A.D. 1400–1640	Clearance cairn
Betula	GrN-28689	440 ± 25	A.D. 1420–1485	Iron production site
Betula	Ua-26808	490 ± 35	A.D. 1330–1470	Hearth
Betula	Ua-26819	640 ± 45	A.D. 1280–1410	Clearance cairn
Betula	Ua-26945	645 ± 40	A.D. 1280–1410	Clearance cairn
Betula	Ua-26960	670 ± 45	A.D. 1270–1410	Clearance cairn
Betula	Ua-26813	715 ± 35	A.D. 1220–1390	House foundation
Betula	Ua-27021	755 ± 35	A.D. 1215–1300	Hearth
Betula	Ua-26874	830 ± 40	A.D. 1060–1290	Iron production site
Betula	Ua-26872	855 ± 40	A.D. 1040–1280	Iron production site

Betula	Ua-27013	1310 ± 35	A.D. 650–780	Not specified
Betula	Ua-27029	8915 ± 65	8270–7820 B.C.	Fire-cracked stones
Calluna	Ua-26838	55 ± 45	A.D. 1680–1960	Clearance cairn
Calluna	Ua-26835	80 ± 40	A.D. 1670–1960	Clearance cairn
Calluna	Ua-26816	200 ± 40	A.D. 1640–1960	Clearance cairn
Calluna	Ua-26841	210 ± 45	A.D. 1520–1960	Clearance cairn
Calluna	Ua-26942	335 ± 45	A.D. 1450–1650	Clearance cairn
Calluna	Ua-27025	6345 ± 50	5470–5150 B.C.	Hearth
Corylus	Ua-26890	1590 ± 45	A.D. 380–600	Hearth
Corylus	Ua-26892	1660 ± 45	A.D. 250–540	Hearth
Corylus	Ua-27022	2885 ± 50	1260–910 B.C.	Hearth
Corylus	Ua-26999	5250 ± 40	4230–3970 B.C.	Hearth
Corylus	Ua-26755	5380 ± 40	4340–4040 B.C.	Hearth
Corylus	Ua-27007	5740 ± 50	4720–4450 B.C.	Cooking pit
Corylus	Ua-27004	5820 ± 45	4790–4540 B.C.	Hearth
Corylus	Ua-26997	5860 ± 45	4850–4590 B.C.	Hearth
Corylus	Ua-27000	5915 ± 50	4920–4680 B.C.	Pit
Corylus	Ua-27005	5935 ± 50	4940–4690 B.C.	Hearth
Corylus	Ua-27003	5950 ± 50	4940¬–4710 B.C.	Cooking pit
Corylus	Ua-27002	5960 ± 50	4950–4710 B.C.	Cooking pit
Fagus	Ua-26 973	80 ± 40	A.D. 1670–1960	Clearance cairn
Fagus	Ua-26964	140 ± 45	A.D. 1660–1960	Clearance cairn
Fagus	GrN-28331	185 ± 15	A.D. 1660–1950	Hearth
Fagus	Ua-26836	245 ± 35	A.D. 1520–1950	Clearance cairn
Fagus	Ua-26864	305 ± 35	A.D. 1480–1660	Charcoal production site
Fagus	Ua-26873	310 ± 40	A.D. 1480–1660	Not specified
Fagus	Ua-26757	335 ± 30	A.D. 1480–1650	Not specified
Fagus	Ua-26969	370 ± 45	A.D. 1440–1640	Clearance cairn
Fagus	Ua-26837	505 ± 40	A.D. 1320–1470	Clearance cairn
Fagus	Ua-26943	635 ± 45	A.D. 1280–1410	Stone wall
Fagus	Ua-26946	685 ± 45	A.D. 1260–1400	Clearance cairn
Fagus	Ua-26866	775 ± 35	A.D. 1210–1295	Water mill
Fagus	Ua-26884	795 ± 40	A.D. 1160–1290	Not specified
Fagus	Ua-26951	870 ± 45	A.D. 1030–1260	Clearance cairn
Fagus	Ua-26962	1000 ± 45	A.D. 900–1170	Stone wall
Fagus	Ua-26963	1135 ± 45	A.D. 770–1000	Clearance cairn
Fagus	Ua-27014	1265 ± 40	A.D. 660–880	Not specified
Fagus	Ua-26966	1295 ± 45	A.D. 650–870	Clearance cairn
Fagus	Ua-26804	1385 ± 30	A.D. 600–690	Clearance cairn
Frangula	Ua-27026	6735 ± 55	5730–5530 B.C.	Hearth
Pinus	Ua-26953	70 ± 45	A.D. 1670–1960	Not specified
Pinus	Ua-26875	75 ± 40	A.D. 1680–1960	Charcoal production site
Pinus	Ua-27024	75 ± 45	A.D. 1670–1960	Pit
Pinus	Ua-26878	80 ± 40	A.D. 1670–1960	Charcoal production site
Pinus	Ua-26867	85 ± 35	A.D. 1680–1960	Water mill
Pinus	Ua-26955	90 ± 45	A.D. 1670–1960	Not specified
Pinus	Ua-26954	95 ± 40	A.D. 1670–1960	Not specified
Pinus	Ua-26828	105 ± 30	A.D. 1670–1960	Charcoal production site
Pinus	Ua-26882	110 ± 35	A.D. 1670–1960	Hearth
Pinus	Ua-26840	130 ± 35	A.D. 1670–1960	Clearance cairn
Pinus	Ua-26834	135 ± 35	A.D. 1670–1960	Clearance cairn
Pinus	Ua-26956	140 ± 45	A.D. 1660–1960	Not specified
Pinus	Ua-26865	150 ± 40	A.D. 1660–1960	Water mill
Pinus	Ua-26876	160 ± 40	A.D. 1660–1960	Not specified
Pinus	Ua-26833	175 ± 30	A.D. 1650–1960	Charcoal production site
Pinus	Ua-26958	195 ± 45	A.D. 1640–1960	Not specified
Pinus	Ua-26832	195 ± 30	A.D. 1640–1950	Charcoal production site
Pinus	Ua-26881	200 ± 40	A.D. 1640–1960	Charcoal production site

Pinus	Ua-26831	205 ± 30	A.D. 1640–1950	Charcoal production site
Pinus	Ua-26830	205 ± 30	A.D. 1640–1950	Stone packing
Pinus	Ua-26829	225 ± 30	A.D. 1630–1950	Stone packing
Pinus	Ua-26877	225 ± 40	A.D. 1520–1960	Not specified
Pinus	Ua-26812	270 ± 35	A.D. 1490–1800	Post hole
Pinus	Ua-26824	460 ± 30	A.D. 1410–1480	Clearance cairn
Pinus	Ua-26825	495 ± 40	A.D. 1320–1480	Clearance cairn
Pinus	Ua-26817	730 ± 40	A.D. 1210–1390	Clearance cairn
Pinus	Ua-26950	745 ± 45	A.D. 1190–1390	Tar pit
Pinus	Ua-26814	755 ± 35	A.D. 1215–1300	House foundation
Pinus	Ua-26948	780 ± 45	A.D. 1160–1300	Charcoal production site
Pinus	Ua-26949	790 ± 45	A.D. 1160–1300	Charcoal production site
Pinus	Ua-26809	875 ± 40	A.D. 1030–1260	Hearth
Pinus	Ua-27033	930 ± 45	A.D. 1020–1220	Pit
Pinus	Ua-26947	1030 ± 45	A.D. 890–1160	Hearth
Pinus	Ua-26820	1115 ± 45	A.D. 780–1020	Clearance cairn
Pinus	Ua-26827	1115 ± 35	A.D. 780–1020	Clearance cairn
Pinus	Ua-27027	2125 ± 40	360–40 B.C.	Hearth
Pinus	Ua-26754	5280 ± 40	4230–3980 B.C.	Pit
Pinus	Ua-27009	6605 ± 50	5630–5470 B.C.	Cooking pit
Pinus	Ua-27017	7155 ± 75	6210–5840 B.C.	Pit
Pinus	Ua-27020	7175 ± 55	6210–5910 B.C.	Hearth
Pinus	Ua-27030	9370 ± 45	8790–8470 B.C.	Fire-cracked stones
Pomoideae	Ua-26895	1250 ± 45	A.D. 670–890	Hearth
Pomoideae	Ua-26891	1665 ± 45	A.D. 250–540	Pit
Quercus	Ua-26 972	35 ± 45	A.D. 1680–1960	Clearance cairn
Quercus	Ua-26970	65 ± 45	A.D. 1670–1960	Clearance cairn
Quercus	Ua-26971	970 ± 45	A.D. 980–1190	Clearance cairn
Quercus	Ua-26888	1110 ± 40	A.D. 780–1020	Hearth
Quercus	Ua-26968	1200 ± 45	A.D. 680–970	Clearance cairn
Quercus	Ua-26889	1430 ± 40	A.D. 540–670	Hearth
Quercus	Ua-27019	2020 ± 45	170 B.C.–A.D. 80	Hearth
Quercus	Ua-27001	2030 ± 40	170 B.C.–A.D. 70	Cooking pit
Quercus	Ua-26894	2420 ± 50	770–390 B.C.	Hearth
Quercus	Ua-26896	2535 ± 51	810–410 B.C.	Hearth
Quercus	Ua-27015	3035 ± 52	1420–1120 B.C.	Cooking pit
Quercus	Ua-26811	4400 ± 35	3270–2910 B.C.	House foundation
Quercus	Ua-26805	4855 ± 45	3730–3520 B.C.	Pit
Quercus	Ua-26887	5270 ± 45	4230–3970 B.C.	Pit
Quercus	Ua-27010	5970 ± 55	5000–4710 B.C.	Cooking pit
Salix	Ua-26810	2095 ± 40	350 B.C.–A.D. 10	Hearth
Salix	Ua-26821	4390 ± 40	3270–2900 B.C.	House foundation
Salix	Ua-27028	7855 ± 65	7050–6500 B.C.	Hearth
Tilia	Ua-26897	1610 ± 54	A.D. 260–570	Hearth
Tilia	Ua-26967	1955 ± 45	60 B.C.–A.D. 140	Clearance cairn
Tilia	Ua-26815	4350 ± 55	3280–2870 B.C.	Clearance cairn

APPENDIX 3: PLANT NAMES

· · · · · · · · ·

Below is a list in alphabetical order of all the English plant names that are mentioned in the book together with their Latin and Swedish equivalents.

Table 14. Plant names.

English	Latin	Swedish
alder	Alnus glutinosa	klibbal
alder buckthorn	Frangula alnus	brakved
aspen	Populus tremula	asp
barley	Hordeum vulgare	korn
bean family	Fabaceae	ärtväxter
bedstraw	Galium sp.	måror
beech	Fagus sylvatica	bok
bilberry	Vaccinium myrtillus	blåbär
birch	Betula sp.	björk
blackthorn	Prunus spinosa	slån
bog asphodel	Narthecium ossifragum	myrlilja
bog moss	Sphagnum sp.	vitmossa
bog-myrtle	Myrica gale	pors
bread wheat	Triticum aestivum	brödvete
buckwheat	Fagopyrum esculentum	bovete
buttercups	Ranunculus sp.	smörblommor
carrot family	Apiaceae	flockblomstriga växter
cereals	Cerealia	sädesslag
cherry	Prunus sp.	körsbär
cincuefoil	Potentilla sp.	fingerörter
clover	Trifolium sp.	klöver
colt's-foot	Tussilago farfara	hästhov
common knapweed	Centaurea nigra	svartklint
common oak	Quercus robur	ek
couch	Elytrigia repens	kvickrot
cow-wheat	Melampyrum sp.	kovaller
dandelions	Taraxacum sp.	maskrosor
docks	Rumex sp.	skräppor
downy birch	Betula pubescens	glasbjörk
dropwort	Filipendula vulgaris	brudbröd
einkorn	Trticum monococcum	enkorn
eyebrights	Euphrasia sp.	ögontröst
field pea	Pisum sativum	foderärt
flax	Linum usitatissimum	lin
floating sweet-grass	Glyceria fluitans	mannagräs
globeflower	Trollius europaeus	smörbollar
goosefoot family	Chenopodiaceae	mållväxter
grass family	Poaceae	gräs

groundsel	Senecio sp.	korsörter
hare's-tail cottongrass	Eriophorum vaginatum	tuvull
hawk's-beards	Crepis sp.	fibblor
hawthorn	Crataegus sp.	hagtornar
hazel	Corylus avellana	hassel
heather	Calluna vulgaris	ljung
hemp	Cannabis sativa	hampa
hop	Humulus lupulus	humle
hornbeam	Carpinus betulus	avenbok
horse bean	Vicia faba	hästböna
hulled barley	Hordeum vulgare var. vulgare	skalkorn
juniper	Juniperus communis	en
knotgrass	Polygonum aviculare	trampört
Labrador-tea	Ledum palustre	skvattram
lilac	Syringa vulgaris	syren
lime	Tilia sp.	lind
lingberry	Vaccinium vitis-idaea	lingon
lyme-grass	Leymus arenarius	strandråg
maple	Acer platanoides	lönn
meadowsweet	Filipendula ulmaria	älggräs
mugwort	Artemisia vulgaris	gråbo
nettles	Urtica sp.	nässlor
Norway spruce	Picea abies	gran
oak	Quercus sp.	ek
oat	Avena sativa	havre
pine	Pinus sp.	tall
potato	Solanum tuberosum	potatis
purple moor-grass	Molinia caerulea	blåtåtel
ribwort plantain	Plantago lanceolata	svartkämpar
rose family	Rosaceae	rosväxter
rowen	Sorbus aucuparia	rönn
rye	Secale cereale	råg
Scots pine	Pinus sylvestris	tall
sedges	Carex sp.	starrar
sessile oak	Quercus petraea	bergek
sheep's-fescue	Festuca ovina	fårsvingel
silver birch	Betula pendula	vårtbjörk
small-leaved lime	Tilia cordata	lind
sorrel	Rumex acetosa/acetosella	ängssyra, bergssyra
spruce	Picea sp.	gran
thistles	Circium sp.	tistlar
tormentil	Potentilla erecta	blodrot
turnip	Brassica rapa	rova
wavy hair-grass	Deschampsia flexuosa	kruståtel
wheat	Triticum sp.	vete
wild-oat	Avena fluitans	flyghavre
willow	Salix sp.	vide, sälg, pil
wormwood	Artemisia sp.	malörter
wych elm	Ulmus glabra	alm
yellow-rattle	Rhinanthus sp.	ängsskallra, höskallra